Landscape with Technology

Landscape with Technology

Essays in honour of L.T.C. Rolt
by Rolt Fellows and members of the
History of Technology Seminar
at the University of Bath

edited by
Professor R. Angus Buchanan

Millstream Books

In Memoriam
L.T.C Rolt
(1910-1974)
with affection and gratitude

Tom Rolt (photograph by Eric de Maré)

First published in 2011 by Millstream Books, 18 The Tyning, Bath BA2 6AL

Set in Bembo and printed in Great Britain by The Short Run Press, Exeter

ISBN: 978 0 948975 92 9

© *David Ashford, Mike Bone, Brenda J. Buchanan, R. Angus Buchanan, Keith Falconer, Robin Morris, Peter Stokes, Geoff Wallis & Owen Ward 2011*

This book is printed on paper certified by the Forest Stewardship Council

All rights reserved. No part of this publication may be reproduced, stored in a retrieval system, or transmitted in any form or by any means electronic, mechanical, photocopying, recording or otherwise, without the prior permission of the authors.

Contents

	Contributors		6
	Introduction	*Angus Buchanan*	7
1	Industrial World Heritage Sites: from icons to landscapes	*Keith Falconer*	10
2	The Birth of British Gunpowder Engineering Overseas: the case of the Mole at Tangier, 1661-1684	*Brenda Buchanan*	22
3	The Lost Distilleries of Bristol and Bath, 1775-1815	*Mike Bone*	43
4	Managing a West Indian Sugar Estate: John Pinney and the Island of Nevis	*Owen Ward*	59
5	James Nasmyth: Engineering Astronomer	*Angus Buchanan*	70
6	Engineering Education in the Age of Microelectronics	*Robin Morris*	81
7	Testing Times: Aerospace and Historic Engines	*Peter Stokes*	88
8	The New Great Space Race	*David Ashford*	100
9	Working Historic Machinery – can it be safe?	*Geoff Wallis*	108
	Index		124
	Postscript: The Rolt Fellowship		128

Contributors

Ashford, David, BSc (Aeronautical Engineering): Rolt Fellow (1997); director of Bristol Spaceplanes Ltd; author of *Spaceflight Revolution* (London: Imperial College Press, 2002).

Bone, Michael, BA, MEd: Rolt Fellow (2005); academic background in history, education and management; past chairman of the Association for Industrial Archaeology (AIA) and Bristol Industrial Archaeological Society (BIAS); member of a number of other local and national heritage committees. author of 'The Rise and Fall of Bath's Breweries: 1736-1960' in *Bath History*, Vol. VIII (Bath: Millstream Books, 2000)

Buchanan, Brenda J., BSc(Econ), PhD, FSA: Visiting Fellow, University of Bath; editor of *Gunpowder: The History of an International Technology* (Bath: Bath University Press, 1996) and *Gunpowder, Explosives and the State: A Technological History* (Aldershot: Ashgate, 2006).

Buchanan, R. Angus, OBE, MA, PhD, FSA, FRHistS: Emeritus Professor of the History of Technology, University of Bath, and Honorary Director of the Centre for the History of Technology; author of *Brunel: The Life and Times of I.K. Brunel* (London: Hambledon & London, 2002).

Falconer, Keith A., MA, FSA: Head of Industrial Archaeology, English Heritage; Visiting Fellow, University of Bath; author of *Guide to England's Industrial Heritage* (London: Batsford, 1980) and (with John Cattell) of *Swindon: Legacy of a Railway Town* (London: HMSO, 1995).

Morris, P. Robin, BA, MPhil, PhD, CEng, MIEE, MIEEE: Rolt Fellow (1990); senior lecturer in Electronic Engineering, Southampton Institute of Higher Education, 1970-1988; author of *A History of the World Semiconductor Industry* (Exeter: Peter Peregrinus, 1990).

Stokes, Peter R., EurIng, CEng, FIMechE, MRAeS: Rolt Fellow (1997); engineering consultant; facility designer and tester of aircraft engines.

Wallis, Geoffrey J.O., EurIng, CEng, MIMechE, FRSA: Rolt Fellow (2005); founder, chairman and managing director of Dorothea Restorations Ltd; now an independent engineering consultant and contractor on the conservation of historic machinery.

Ward, Owen, BA: Visiting Fellow, University of Bath; retired university administrator; author of papers on mill technology and on industrial sites in the Bristol/Bath area.

Introduction

Angus Buchanan

L.T.C. Rolt, pioneering conservationist and engineering historian, was born in 1910. When he died, in 1974, he had just been awarded an honorary degree by the University of Bath, the Senate of which decided that it would be appropriate to honour his memory by creating a Research Fellowship in his name. The Rolt Fellowship was accordingly set up, financed by a fund raised by an appeal to friends and admirers of Tom Rolt, and dedicated to the idea of encouraging mature professional men and women to prepare for publication research in the history of technology. The scheme has been administered from the outset by the Centre for the History of Technology at the University of Bath.

The scheme was established in 1978, when the first Fellow, J.H. Boyes, a retired Inspector of Factories, was selected, and since then a total of 14 Fellows has been appointed. The Fellowship provides a small grant to help with research expenses, but the main benefit has been the association with the University as honorary Visiting Research Fellows, using the Library facilities and joining in the activities of the History of Technology Seminar which has been held twice a term for many years by the Centre for the History of Technology. Fellows have maintained their contact with the seminar long after completing their original project, and a valuable camaraderie has developed over the years, enabling members to undertake a number of joint research projects on topics such as engineering disasters and engineering education, both of which have led to useful publications. At present, with six active Rolt Fellows, three Visiting Research Fellows, several local supporters, and the Honorary Director of the Centre, the seminar has continued to sustain research activity in the history of technology at the University, and the centenary of Rolt's birth has persuaded them to commemorate the event with a tribute in the form of a collection of essays by members of the seminar.

Tom Rolt trained as an engineer, serving a complicated apprenticeship first with Bomfords, agricultural engineers at Evesham; then with Kerr, Stuart & Co, locomotive engineers at Stoke-on-Trent; and finally with R.A. Lister & Co, agricultural engineers of Dursley. In 1934 he joined some friends in running a garage and establishing the Vintage Sports Car Club, but five years later he abandoned his engineering practice to pursue a career as a writer while living on a canal narrow boat, the *Cressy*. He and his first wife made a voyage of the Midland canals just before the outbreak of the Second World War – during which the boat was moored and Rolt undertook various official engineering commissions for the government – but the voyage resulted in the publication, in 1944, of *Narrow Boat*, an eloquent evocation of the canal system, then in serious decline, and of the people who worked upon it. The book stimulated an enthusiastic response and led to the formation of the Inland Waterways Association in 1946, for which Rolt served as Secretary for five years. Another phase in the career of Rolt as a crusading conservationist began in 1950, when he became involved in a campaign to re-open the Talyllyn Railway, a narrow-gauge railway in North Wales built to serve a quarry that had closed down, making the line redundant. It was successfully revived as a leisure amenity, staffed largely by amateurs, and set a pattern for numerous similar restorations in many parts of the country, with Rolt's tireless advocacy in books and articles playing a vital part in achieving public support.

In this period Rolt settled with Sonia, his second wife, in what had been his family house at Stanley Pontlarge, near Winchcombe in Gloucestershire, where he lived for the remainder of his life, devoting himself to writing over 40 books and to the many conservationist enterprises in which he became involved. These dealt with canals, railways, and motor cars, and Rolt displayed outstanding skill as a biographer of engineers, writing vivid accounts of the lives of the great nineteenth-century engineers such as *Thomas Telford*, *George and Robert Stephenson*, and *Isambard Kingdom Brunel*. Like all his books, these were written with great lucidity, so that the technical aspects of engineering were presented in a meaningful manner to the non-technical reader, while retaining technical authenticity and stylistic felicity. They enjoyed a wide readership, and did much to encourage a revival of interest in engineers and engineering history in the middle decades of the twentieth century.

His success as a writer brought Tom Rolt many concomitant responsibilities, as he was invited to support the enthusiasts who rallied to his call to save the canals, to revive narrow-gauge railways, to preserve vintage cars and steam engines, to create museums, and generally to stimulate consciousness of the heritage value of technological history and its artefacts. Many national bodies such as the National Trust and the Science Museum sought his advice, the latter in the establishment of the ambitious Railway Museum in York, and he became a Vice-President of the Newcomen Society for the study of the history of engineering and technology, and Chairman of the Council for British Archaeology research committee on industrial archaeology. The formation of the Association for Industrial Archaeology, of which Rolt became the first President, was his last great institutional achievement, and he died on 9 May 1974, less than a year after its foundation.

The essays which are presented here in Rolt's memory cover a wide range of subjects, most of which have been the subject of attention by the Bath seminar in recent years. Although no single theme has been prescribed, they are all of potential interest to historians of technology and fit comfortably under the title *Landscape with Technology*, which echoes the titles of the three volumes of autobiography that Tom Rolt composed towards the end of his career: *Landscape with Machines* (Longman, 1971), *Landscape with Canals* (Allen Lane, 1977), and *Landscape with Figures* (Alan Sutton, 1992). While not referring specifically to the work of Rolt, the contributors to this collection are all deeply conscious of the significant role played by him in industrial history, engineering biography, environmental conservation, and concern for the national and international heritage. It seems appropriate, therefore, to take a title for our mixture of topics which reflects these enthusiasms of Rolt, which we all share. Keith Falconer writes with intimate experience of the development of World Heritage sites, which have rightly included many outstanding technological monuments; and Mike Bone draws on his great experience of regional research on the West Country distilling industry. Then Brenda Buchanan explores the technological significance of the attempt to develop the port of Tangier in the seventeenth century; and Owen Ward discusses the financial and technical problems of running a West Indian estate in the eighteenth century. Angus Buchanan considers a little-known aspect of the career of one of the great nineteenth-century mechanical engineers, who was also an accomplished amateur astronomer. Robin Morris deals with some of the problems of adapting engineering education to the needs of a society in which microelectronics has become an essential part; and Peter Stokes reflects on a career spent between testing aeroplane engines and restoring stationary steam engines. Then David Ashford analyses sharply the fixed mind-sets which have led to an over-reliance on expendable launchers rather than fully reusable spaceplanes in the development of the space programme. Finally, Geoff Wallis gives a carefully considered judgment on the health and safety problems of working historic machinery.

All of us are united in our admiration of the work of Tom Rolt, whose achievements we salute in this centenary tribute.

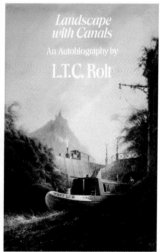

1: Industrial World Heritage Sites: from icons to landscapes

A celebration of L.T.C. Rolt and 'industrial' World Heritage Sites

Keith Falconer

On 25 June 2009 the Pontcysyllte Aqueduct along with an 11-km stretch of the Llangollen Canal was inscribed as a World Heritage Site (WHS) by the World Heritage Committee meeting in Seville. How Tom Rolt would have rejoiced! That the aqueduct should join the Ironbridge Gorge and the Canal du Midi in that Pantheon of Sites would have been a particular pleasure for they all meant so very much to Tom. The Pontcysyllte Aqueduct was the ultimate, but unrealised objective of his honeymoon voyage in July and August 1939 and its crossing was an ambition only to be achieved with great difficulty ten years later. The second part of his autobiography – Landscape with Canals – recounts the trials and tribulations surrounding that achievement and his first exposure to the rare beauty of the Ironbridge Gorge and Coalbrookdale. The Canal du Midi was the venue of his last epic voyages in spring 1971 and September 1972 and the subject of one of his finest books – From Sea to Sea – published shortly before his death two years later. This tribute to Tom takes industrial World Heritage Sites as its theme, outlining the changing perceptions of these sites and how the emphasis on single iconic sites has evolved to an appreciation of landscapes. An appreciation that Tom himself pioneered more than half a century ago.

Firstly let us turn the clock back some 30 years. The first industrial site to be inscribed on the World Heritage List was the Weiliczka salt mine in Poland in 1978. This mine, with its 300 kms of galleries, vast wood-propped caverns, carved shrines and over eight centuries of mining operation, had been attracting visitors since the fifteenth century and is very much an iconic cultural monument though it also preserves dramatic technological evidence. This was followed in the 1980s by the mining settlement of Roros in Norway with its well-preserved wooden vernacular housing, and then by further iconic monuments – in 1982 by Ledoux's grandiose Royal Saltworks of Arc-et-Senans in France and in 1984 by the Statue of Liberty, New York.

Then in 1986 the Ironbridge Gorge in England was inscribed. With this nomination the UK broke new ground in seeking recognition for an extensive industrial landscape rather than a single site. This was very much in keeping with Tom's views on industrial heritage. Regularly staying in the Valley Hotel at Coalbrookdale during his wartime spell working as an 'Isolated Technical Assistant' for the Ministry of Supply, he would explore the gorge on foot in the evenings, inspired by the pictures that hung on the walls of the hotel bar. The three 1788 coloured engravings of George Robertson's scenes depicted not only the famous bridge but also furnaces and smelting houses at Broseley with their 'breathing fire and foul smoke … resembling the mouth of hell'. Tom was able to share the

George Robertson's print of 1788 which Tom Rolt would see during his sojourns in the Valley Hotel, Coalbrookdale. (Ironbridge Gorge Museum Trust)

fascination of artists drawn to the gorge to express in paint the dramatic contrast between 'the fuming, flaming clangour of the ironworks and their idyllic setting of wooded hills and streams'. On these walks 'everywhere I was reminded of the fierce activity of former days, and every stick and stone of the place seemed to have absorbed something of its white hot violence'.[1] As he puts it, the landscape and the great black semi-circle of Darby's bridge spoke more eloquently than any history book. What better justification for a World Heritage Site. But how different from Tom's days when the silent furnaces were overgrown and unloved. Thirty-five historic sites within the World Heritage Site have benefited from more than £30 million pounds of public investment and the ten museums run by the Ironbridge Gorge Museum Trust attract over half a million visitors a year.[2]

Emboldened by the inscription of the Ironbridge Gorge, over the next decade further extensive industrial sites were nominated, including the metal-mining towns of Rammelsberg and Goslar in Germany, and Guanajuanto in Mexico; Engelsberg ironworks in Sweden; Volklingen ironworks in Germany; and Crespi D'Alba textile mills and settlement in Italy. Then in 1996 the Canal du Midi was inscribed. This would have thrilled Tom though not surprised him. Though a self-professed stay-at-home anglophile, Tom's lifelong interest in canals inspired in him 'a strong desire to see for myself the Canal du Midi, that archetypal engineering work which … set an example not only to England, but to the world'.

The Canal du Midi – a typical lock and lock-keeper's house. (author's collection, 2006)

Tom, unlike the young Duke of Bridgewater who only saw the canal from the bank, explored the canal by boat in two voyages – eastwards in 1971 from Castelnaudary to the Étang de Thau on the Mediterranean and then the following September from Agde on the Mediterranean all the way to Toulouse and thence down the Canal Latéral à la Garonne to Bordeaux. The chronicle of those voyages has inspired generations of canal enthusiasts ever since. Further canal monuments were to be inscribed – the four Canal du Centre lifts in Belgium in 1998; the Rideau Canal in Canada in 2007; and the Pontcysyllte section of the Llangollen Canal of which more anon.

The Thatcher years were a fallow period as far as British World Heritage Sites were concerned – the government set its heart against UNESCO initiatives and no nominations were forthcoming. Elsewhere there were no such reservations and a steady stream of industrial sites was inscribed – the Verla board mill in Finland; the San Leucio silk mill Caserta near Naples; a cluster of windmills at Kinderdijk and the Wouda steam pumping station in Holland; and the Karlskrona naval port and dockyard in Sweden.

In 1997, to mark the 25th anniversary of the signing of the World Heritage Convention, English Heritage hosted a celebratory conference in London. The conference flagged up the issues of certain cultural landscapes, and the industrial heritage in particular, being under-represented on the World Heritage List in the past. In Britain, the implications

Bas-relief at the junction of the Canal du Midi with the Canal Latéral, Toulouse. (author's collection, 2006)

of this shift in UNESCO thinking were immediate. The joint committee of heritage agencies overseeing the production of a new UK *Tentative List* heeded these promptings of the World Heritage Convention to widen the range of sites and responded vigorously to the Convention's singling out of the industrial heritage for particular attention. The consequences for the industrial heritage were immense – no less than 11 of the 21 UK mainland sites promoted in the 1999 UK *Tentative List* are industrial – seven in England and two each in Scotland and Wales.

The English selection of industrial sites was co-ordinated in 1998 by English Heritage's Industrial Archaeology Panel – the successor to the CBA Panel chaired by Tom Rolt some 20 years before. Over 100 suggestions, made by public consultation, were narrowed down to a short list of 30 to be examined more closely and then just seven were developed. The panel adopted the innovative approach of seeking to identify themes and to illustrate these by landscape designations, rather than focusing on single pre-eminent sites. It was this emphasis on themes illustrating the unique British contribution to world industrialisation that was developed and subsequently articulated in the *Tentative List*. In the words of the Secretary of State for Culture Media and Sport, this reflected a concern 'to advance the concept of World Heritage beyond the monumental and architectural into areas of relevance to all humanity'. He added 'the inception and process of industrialisation ... has changed and moulded the way in which all the peoples of the world now live. That process began here in Britain and it is right that it should be marked more prominently in the World Heritage List.'³

The sites selected had to ensure balanced representation of Britain's contribution to the world's heritage taking into account sites already on the *List*. The intellectual case for the selection of themes and the selection of sites themselves would have to be rigorous and robust to withstand the vigorous examination and counterclaim that would inevitably follow from within the UK. It was appreciated that the emphasis on themed landscapes presented conceptual problems as a landscape is the sum product of a great many cultural factors operating across time and topography but nevertheless it was felt that discrete landscapes could be meaningfully characterised by dominant activities. The themes selected were:

J.M.W.Turner's painting Rain, Steam and Speed – The Great Western Railway *on the front cover of the 1999 UK* Tentative List of World Heritage Sites. *The inclusion of the GWR in the list chimes with Tom Rolt's championship of Brunel and his inspired construction of that railway.*

- the industrialisation of processing and manufacture as exemplified by the textile industry in the Derwent Valley, Ancoats and Saltaire in England and by New Lanark in Scotland;
- pioneer developments in inland transport illustrated in the Manchester region by the structures of the earliest industrial canals including the Bridgewater Canal and passenger railways exemplified by the Manchester terminus of the Liverpool & Manchester Railway;
- virtuosity in civil engineering demonstrated by the Great Western Railway from London to Bristol and by the Pontcysyllte Aqueduct in Wales and the Forth Bridge in Scotland;
- developments in eighteenth- and nineteenth-century mining which created such distinctive landscapes in Britain and, by the export of the technology, similar landscapes across the world. In England hard-rock, deep mining techniques are represented by several distinctive Cornish landscapes while coal-mining (with iron-making) is represented by the landscape around Blaenavon in Wales;
- the global maritime and naval influences encapsulated by Liverpool's historic waterfront and related commercial and institutional buildings and Chatham Royal Naval Dockyard.

In establishing the applicability of the concept of landscape designation in an industrial context the panel had to conduct its own research, consult with national and local interests and expertise and undertake site visits to establish the extent and authenticity of the surviving remains. The UK *Tentative List* was published in 1999 and attracted worldwide attention as regards the industrial heritage where it was seen as setting an international agenda. Meanwhile the Semmering Railway in Austria and Darjeeling Himalayan Railway had been inscribed in 1998 and 1999 respectively. The Darjeeling, opened in 1881, is still the most outstanding example of a hill passenger railway.

To return to the British scene: the Blaenavon bid was the first to be considered and this nomination built on the success of the earlier Ironbridge Gorge inscription, taking the concept of landscape a step further. It justifiably claimed that the Blaenavon industrial landscape presents a large number of individual monuments of outstanding value within the context of a rich and continuous relict landscape, powerfully evocative of the Industrial Revolution. It is one of the prime areas in the world where the full social, economic and technological process of industrialisation through coal and iron production can be studied and understood. Blaenavon was inscribed as a World Heritage Site in 2000. The following year the World Heritage Committee inscribed three linked sites from the UK to represent the industrialisation of processing and manufacture as exemplified by the textile industry – the Derwent Valley Mills, New Lanark and Saltaire. These focused on the theme of the development of the textile factory system, witnessing innovations in the harnessing of water and steam power to textile processes and on the transformation of the landscape by the scale and structure of their manufacturing buildings and the planned housing of the labour force.

Subsequently, the Liverpool Maritime City and the Cornish Mining Industry World Heritage Sites have been inscribed. The former is the supreme example of a commercial port developed at the time of Britain's greatest global influence while the latter takes the UK themed landscape approach to its extreme. The Cornish Mining WHS comprises ten detached areas which collectively represent the many facets of metal mining in the south-western peninsula of England and have significance far beyond Britain itself. Though generally technologically based, the ten discrete areas differ greatly from one another in character and each has a core area that is quite distinct but complements the other core areas. They embrace, for example, the landscape around Camborne/Redruth which, focused on the Great Flat Lode, is the archetypal Cornish landscape of engine houses and mine complexes served by urban settlements of terraced houses with chapels, mining exchanges and engineering works. This area contrasts sharply with the area of coastal mining at St Just where undersea mining was pioneered at mines such as Botallack, Levant and Geevor, and where the remains present some of the most dramatic mining landscapes in the world. These are complemented by the upland mining landscape of Caradon on Bodmin Moor and the Tamar valley, with its smelt works, arsenic calciners and mineral ports.

Pontcysyllte Aqueduct, North Wales; tinted engraving by George Pickering. (Ironbridge Gorge Museum Trust)

The bid for the Pontcysyllte Aqueduct, though primarily focused on Telford's iconic aqueduct – a pioneering cast-iron structure and the highest aqueduct then built, also included the 11-km section of heavily engineered canal either side. It therefore embraced Telford's earlier Chirk Aqueduct and the adjacent industrial landscape of early tramways. The condition of this magnificent stretch of canal is very different from that described by Tom Rolt when he made his tortuous and abortive attempt to reach it in 1947, only to return triumphantly in 1949. The water regulation flow meter at Llantisilio below Telford's Horseshoe Weir, under construction in 1947, was by then operational allowing much more water into the Llangollen (or Welsh) Canal. Cruising west from Chirk Bank high on the hillside, Rolt's description of his first glimpse of the Ceriog Valley and Chirk Aqueduct is a thrill still to be experienced today and, as Tom was to assert, there is no more dramatic entry from England into Wales than by Telford's Chirk Aqueduct.[4]

The shape of a new UK *Tentative List of World Heritage Sites* was under discussion in 2009 and hence the composition of nominations was uncertain. Further 'industrial' bids had been partially developed for Greater Manchester, the Great Western Railway, and Chatham Royal Naval Dockyard but any detailed elaboration awaits the review of the UK *Tentative List*. The Greater Manchester bid focuses on Manchester as the archetypal city of the Industrial Revolution with Britain's first industrial canal, first mainline inter-city passenger railway, and the creation of the first industrial suburb based on canals and steam power. The Great Western bid would celebrate virtuosity in civil and mechanical engineering and comprises the Great Western Railway from London to Bristol; Brunel's railway engineering

Chirk Aqueduct, Llangollen Canal – 'no more dramatic entry into Wales'. (author's collection, 2007)

works and railway village at Swindon; and, to recognise Britain's contribution to maritime engineering, the Great Western Dock with the s.s. *Great Britain*. The Chatham bid, focusing on the preserved sites comprising Historic Chatham, would represent Britain's outstanding contribution to naval development in the eighteenth and early nineteenth centuries.

Meanwhile, elsewhere, nominations were becoming more adventurous in date as well as extent, as evidenced by the Great Copper Mountain, at Falun in Sweden, where ore had been extracted for seven centuries; the vast Zollverein coal mine complex in Germany dating from the twentieth century; and the Varberg radio station representing twentieth-century communications. The first inscriptions from South America – two mining complexes in Chile – for once celebrate monumental human endurance achievements rather than outstanding physical monuments. Interestingly, they represent extreme mining conditions – for saltpetre at Humberstone and Santa Laura, and for copper at Sewell. Otherwise, transport sites were again to the fore. The aftermath of the Balkan conflict witnessed inscriptions celebrating reconciliation with the rebuilding of the bridge at Mostar and Muslim engineering with the Visegrad Bridge, while the Rideau Canal in Canada and the Rhaetian Railway in Switzerland were inscribed in 2007 and 2008 respectively. At this time concern for distinctive agricultural landscapes and an UNESCO World Heritage Centre thematic expert report on vineyard cultural landscapes led to half a dozen such landscapes being inscribed, including the coffee plantations in Cuba; the port-producing Douro region in Portugal; the Tokaj wine region in Hungary; the Pico vineyards in the Azores; the Lavaux vineyards in Switzerland; and the Tequila landscape in Mexico. Though very different from the English rural landscapes so celebrated in Tom Rolt's autobiographies, the recent recognition of agricultural landscapes as of outstanding universal value would have gratified Tom who was so very aware of the amount of human endeavour that such landscapes demanded.

Finally, it is most fitting that a third important element in Tom's life should accord with World Heritage recognition. In 2005 the Darjeeling Himalayan Railway was joined by the Nilgiri Rack Railway, a 46-km-long metre-gauge railway to form the Indian Mountain Railways serial World Heritage Site. The Nilgiri scales some 1,900 metres and opened in 1908, having taken 17 years to build. In 2008 the 96-km Kalka Shimla Railway was inscribed as a third element of the serial Site and the tiny Matheran Hill Railway is currently being considered as a fourth component. This recognition of the universal significance of narrow-gauge railways would have again greatly pleased Tom as such railways were to play such a rewarding part in his later life, as evidenced in the third part of his autobiography *Landscapes with Figures*.

Tom Rolt (fourth from right) with friends at the Talyllyn Railway. (Ironbridge Gorge Museum Trust)

The last 30 years have indeed witnessed a huge shift in perception of World Heritage Sites. The preponderance of iconic monuments on the early world lists has been leavened by the introduction of many varied but outstanding landscapes. Transport monuments have contributed greatly to this shift and Tom would not have minded that so many of the canals and railways inscribed as World Heritage Sites were overseas and relatively few in Britain. More than 35 years ago he railed against 'the chauvinistic belief that British engineering skill was mysteriously and innately superior to that of any other nation'[5] but he would, however, perhaps be slightly surprised and saddened that no British railway site has yet been inscribed on its own account. I.K. Brunel's Great Western Railway, the design and construction of which features so prominently in Rolt's classic biography of Brunel, is one of the entries in the *Tentative List* that has not yet been inscribed as a World Heritage Site.[6] If it were, the s.s. *Great Britain* would be included in the World Heritage Site as part of Brunel's vision of an integrated transport system stretching from London to New York.[7]

(above) Brunel's Great Western Railway, Box Tunnel Western Portal; engraving by J.C. Bourne, 1841. (Ironbridge Gorge Museum Trust)

(right) s.s. Great Britain, in the Great Western Dock, Bristol. Brunel's pioneering steamship returned to Bristol in 1970 and has been magnificently restored. (s.s. Great Britain Trust)

In the post-war years Tom was one of a small band of pioneers in his championing of landscapes throughout his writings. This was finally to be explicitly epitomised in his three-volume autobiography, the last two volumes of which were published posthumously by his widow Sonia Rolt. That landscapes are in the majority in the list of World Heritage Sites celebrating the industrial heritage shown overleaf, shows how perceptive Tom was all those years ago.

World Heritage Sites celebrating industrial heritage
(compiled from UNESCO World Heritage List, http://whc.unseco.org)

Date of Inscription
1978	Weiliczka salt mine, Poland
1980	Roros metal-mining town, Norway
1982	Royal Saltworks of Arc-et-Senans in France, extended in 2009 to include Salins-les-Bains salt works
1984	The Statue of Liberty, New York, USA.
1986	The Ironbridge Gorge, England
1988	Guanajuato town and mines, Mexico
1992	Mines of Rammelsberg and Goslar town, Germany
1993	Engelsberg ironworks, Sweden
1994	Volklingen ironworks, Germany
1995	Crespi D'Alba textile mills and settlement, Italy
1996	The Canal du Midi, France
1996	Verla board mill, Finland
1997	Kinderdijk cluster of windmills, Holland
1997	San Leucio silk mill, Italy
1998	Karlskrona naval port and dockyard
1998	The four Canal du Centre lifts, Belgium
1998	Wouda steam pumping station, Holland
1998	Semmering Railway, Austria
1999	Darjeeling railways; 2005 & 2008 Indian mountain railways
1999	Beemster Polder, Holland
2000	Blaenavon, Wales
2001	Falun mining area of the Great Copper Mountain, Sweden
2001	Zollverein coal mine, Germany
2001	Derwent Valley mills and settlements, England
2001	New Lanark mills and settlement, Scotland
2001	Saltaire mills and settlement, England
2004	Liverpool Maritime City, England
2004	Varberg radio station, Sweden
2005	Mostar Bridge, Bosnia
2005	Plantin-Moretus workshops, Belgium
2005	Humberstone and Santa Laura saltpetre works, Chile
2006	Cornish mining landscapes, England
2006	Sewell mining town, Chile
2007	Visegrad Bridge, Bosnia
2007	Rideau Canal, Canada
2007	Iwami Ginzan silver mines, Japan

2008	Rhaetian Railway, Italy and Switzerland
2009	Pontcysyllte Aqueduct and canal, Wales
2009	La Chaux-de-Fonds/Le Locle watch-making town, Switzerland

Agricultural and viticulture

2000	Coffee plantations, South-east Cuba
2001	Douro region, Portugal
2002	Tokaj wine region, Hungary
2004	Pico vineyards, Azores
2006	Tequila landscape, Mexico
2007	Lavaux vineyards, Switzerland

Notes

1. L.T.C. Rolt, *Landscape with Canals*, p 55 (London: Allen Lane, 1977).
2. Ironbridge Gorge Museum Trust personal communication, 2009.
3. DCMS, *United Kingdom Tentative List of World Heritage Sites*, p.4 (London: HMSO, 1999). J.M.W. Turner's painting *Rain, Steam and Speed – The Great Western Railway* (1844) was chosen by the present author as the cover of the *Tentative List* to symbolise the UK's contribution to global industrialisation.
4. L.T.C. Rolt, *Landscape with Canals*, p.150.
5. L.T.C. Rolt, *From Sea to Sea, The Canal du Midi*, p.4 (London: Allen Lane, 1973).
6. L.T.C. Rolt, *Isambard Kingdom Brunel* (Harlow: Longman, 1957)
7. Keith Falconer (ed.), *The Great Western World Heritage Site: The genesis of modern transport* (Swindon: English Heritage, 2006).

2: The Birth of British Gunpowder Engineering Overseas: The case of the Mole at Tangier, 1661-1684*

Brenda Buchanan

The road to Tangier begins in a country house a few miles north of Bath, set in a deer park from which the name Dyrham was derived. The house was re-built and re-furbished by William Blathwayt after his marriage in 1686 to Mary Wynter, whose family had owned the estate for a hundred years.[1] He was an ambitious civil servant, doing well in royal service with the assistance of his uncle, Thomas Povey.[2] The new east wing built by Blathwayt shows his preference for the Baroque style of the several north European cities in which he served as a diplomat, rather than the Italian Palladian style then becoming popular.[3] Some 40 pieces of fine Delftware survive to recall his time in the Low Countries, his command of the Dutch language and his later service to William and Mary of Orange when they came to the throne in 1689 and popularised such items. Blathwayt's time in the Plantations Office in Whitehall is reflected in the woods from Virginia – the panelling of black walnut in the Diogenes Room and the red cedar and walnut in the fine sweep of the Cedar Staircase.[4] Ascending this staircase to a small landing presents the most surprising 'tribute' of all – a large (68in x 96in) painting by a contemporary but unknown artist, showing the royal possession of Tangier with its fine Mole (Figure 1, pp.24-25). But this vivid depiction of a 'Landscape with Technology' is no scene of imperial triumph for as the painting shows, a large army of Moors was then laying siege to the port-city from which the English were shortly to withdraw their soldiers and sailors, blowing up not only military fortifications and civilian buildings, but also this fine feat of civil engineering, the Mole.

It is the purpose of this study to explore the ways in which from this retreat a sense of achievement may be salvaged because in the short period of less than 25 years, the successful construction of the Mole brought together several important but underrated and neglected themes. These include the early use of black/gunpowder in the quarrying of building materials; the employment of Pozzolana cement (referred to in the documents as 'tarras', 'tarres' or 'tarris') to set the underwater structures before its general employment for this purpose in England; the awareness by the engineers of similar work undertaken at less-challenging Mediterranean harbours such as Genoa and their links with those working on these projects; and the relationship between this work of civil engineering and the military engineers in Tangier.

Tangier, in Morocco, had become a possession of the English Crown as part of the 1661 marriage settlement of the Portuguese Princess Catherine of Braganza who was to

---

* Earlier versions of this paper were presented in 2009 at a seminar of the Centre for the History of Technology at the University of Bath, and at the Pittsburgh Conference of the American Society for the History of Technology.

become the wife of Charles II. Charles had only recently been restored to the throne in 1660 after the Civil Wars of the 1640s, the execution of his father Charles I in 1649, and the experiment of Cromwell's Commonwealth. He was then a young man of 30, looking for a suitable bride with whom marriage would signify a political as well as a dynastic or family connection. Catherine brought as part of her dowry not only these intangibles and a promise of £800,000, which materialised very slowly, but also two Portuguese outstations: Tangier, in north-west Africa, and Bombay (now Mumbai) on the west coast of India. Both presented a foothold in a new continent for the English, who had so far lagged behind the Portuguese, the Spanish, and the Dutch, in their overseas enterprises. Bombay became a very successful trading station, operated by the East India Company, but Tangier, which had promised so much, was abandoned less than a quarter of a century later.[5]

The acquisition of Tangier was greeted with great enthusiasm and expectation, especially as there was much rich trade to be had in the Mediterranean. Before the transatlantic and far eastern routes were fully opened up it constituted the most profitable market in the world, and Tangier guarded the entrance to it. Here then was the opportunity to protect shipping, especially from the Barbary corsairs of North Africa who were active not only in the Mediterranean but also nearer home at the mouth of the Thames Estuary and in the Bristol Channel.[6] In the longer term this base would offer the chance to build up the army, because after Cromwell and the New Model Army there was no appetite for a standing army in England; and it would provide facilities for the navy because the fleet had then no base outside home waters and as the constructor Sir Hugh Cholmley observed, the Mole was 1,200 miles from any English possession.[7] Charles II also had far-reaching personal and national ambitions for the profits and opportunities that would come from setting up such a port-city, noting in a Royal Proclamation of 1662 that 'the welfare of Our good subjects depends very much upon the safety and improvement of Trade'.[8] There was also an awareness that Tangier could become the focus of an empire closer to home than India or North America, perhaps leading to settlement and dominion in Africa.

First, however, the physical problems had to be recognised and solved. Tangier was located on the exposed north-west tip of Africa, with heavy seas, fierce winds, Atlantic storms, deep waters, and substantial tidal movements on the west, and exposure to the strong Levantine winds from the east. The existing harbour was so inadequate that the construction of a mole or breakwater was seen as central to the whole enterprise. This major undertaking was to be the first great harbour work in deep tidal waters, and possibly the first example of gunpowder engineering by a European country outside that continent. And although the project was conceived in response to political and military requirements, so that the association with the Board of Ordnance was very strong, the work of construction, especially the methods used for blasting and setting stone, and the organisation of the craftsmen involved, was a matter of civil engineering, with for much of the time a civilian contractor and workmen. The political context for this work

Fig.1. The Siege of Tangier 1683 *(68in x 96in), by an unknown seventeenth-century*

(reproduced by courtesy of the House & Collections Manager, Dyrham Park, the National Trust)

exacerbated these difficulties. Most immediately, the Portuguese living in Tangier were unwilling to leave, but more generally and more seriously, this was a time of upheaval in the history of Morocco. Referred to by Europeans as the Moors, these Arabs of north-west Africa were generally Muslims although not part of the Ottoman Empire spreading along the coast of north Africa and encompassing Algiers, Tunis and Tripoli. There had been attempts at creating Moroccan unity under an Emperor based in Fez, but some leaders such as Ab'd Allah Ghaïlán, who commanded a fighting force based near Tangier, had broken away. To the English he was known informally as 'Gayland' and formally but incorrectly as the 'Emperor and Prince of West Barbary', and it was with him that arrangements regarding, for example, food and pasture land had to be made.[9] But in a more sinister fashion these negotiations often also included firearms and powder, and if these could not be obtained through official channels, there were always English merchants willing to sell to the highest bidder.[10]

In January 1662 the Earl of Sandwich took possession of Tangier on behalf of the King, and picked a suitable site for the Mole.[11] An initial survey was made by Martin Beckman, a military engineer of Swedish origin.[12] Encouraged by favourable reports the Privy Council's Committee for Tangier decided to enter into a contract for the project with three partners: Andrew Rutherford, appointed Governor, soon to become Lord Teviot; Sir John Lawson, the naval commander, who saw this as striking a blow at the rival 'Hollanders' who might otherwise 'give the law to all the trade in the Mediterranean'; and Hugh Cholmley, a country landowner soon to inherit the family title and become Sir Hugh, who was to be the working engineering partner, engaged on the site. The contract was signed in March 1663. By these specifications the contractors were to build the Mole of stone, 30 yards broad at its foundation, and running from the shore out to sea for 400 yards, turning more or less a right angle at the end and running for another 200 yards. An advance of £2,000 was to be made towards a grossly under-estimated cost of £10,000. It was to cost more than £340,000.[13]

It is curious that with so many talented military engineers available, the construction of the Mole was placed in the hands of Hugh Cholmley.[14] Setting aside all thoughts of influence at Court, of which there is no evidence (although Samuel Pepys records in his diary that money had changed hands[15]), it is likely that it went to Sir Hugh because of his experience and that of his family, in improving the harbour at Whitby, a small but busy seaport on the north-east coast of England, lashed by the winds and waves of the North Sea. Of course these harbour works were not altruistic undertakings for the Cholmleys were entrepreneurial gentry, and in addition to a pleasant country house based on monastic foundations acquired from King Henry VIII, they had alum workings and quarries in the hinterland. The former prospered because of the growth of the textile industry, and from the latter (especially the quarries at Aislaby) came large blocks of hard-wearing stone that was much in demand. It made excellent material for piers and breakwaters, and was used not only at Whitby but for similar work elsewhere. Some was even shipped to Tangier. The work on the Whitby mole or pier in the 1650s, done 'on a new plan' involving a substantial 'projection' of some 230 yards from the Scotch Head on the west side of the harbour, was probably the achievement which secured for Sir Hugh a place on the Tangier Committee in December 1662, and the contract of March 1663.[16] Sir Hugh's colleagues in Tangier had the advantage over him in military engineering, but they could not match him in the experience of civil engineering.

An expedition set out for Tangier in June 1663. Hugh Cholmley had with him 'about forty masons, miners and other proper artists and workmen', and Jonas Moore, a skilled surveyor who was very impressed by the possibilities at Tangier. Back in London in September 1663, Moore reported that with the major undertaking of a defensible harbour this was 'likely to be the most considerable place the King of England hath in the world'. He may have made a second visit for there is a reference in May 1664 to a map 'new drawn' by him, showing the city of Tangier and the Straits of Gibraltar, implying an understanding of the strategic and not just the trading importance of the place. Jonas Moore was to become Surveyor General of the Board of Ordnance in 1669.[17]

As befits a Royal enterprise the whole operation was well-documented through the reports and discussions noted in official state papers, especially those of the Tangier Committee of the Privy Council. There are also excellent sets of illustrations, prepared by artists sent out to record progress for the King or by military engineers of the Board of Ordnance.[18] But most particularly we are fortunate because Sir Hugh Cholmley kept a good and careful account of how this work of engineering was undertaken, paying special attention to the use of gunpowder in quarrying, in a detailed description that is both rare and early.[19] As a classically-educated Cambridge man, Sir Hugh, then in his early thirties, began his account by reflecting on the Latin origins of the term 'mole' meaning a mass, in this case of stones. The terms 'pier' and 'cobb' were used for similar features he observed, and it may be noted that the wall of the small harbour at Lyme Regis on the English south coast has been known as 'The Cobb' since the twelfth century. But Sir Hugh's focus was on earlier moles in the Mediterranean, and he refers to that at Ostia near Rome, described by Pliny, and the more recent case of Genoa where work was undertaken in the 1630s. These are significant examples because they illustrate the different methods by which a mole may be constructed. In both cases the great quantities of stone required could be carried by ship, but at Ostia, this was cast out in a 'confused manner … so that at last its [the mole's] rocky back appeared above water', whilst at Genoa great chests holding the stones were lowered into the water until the desired level and platform was reached. But unlike the Mediterranean proper where there was 'no regular motion', matters at Tangier were complicated by the great tidal range of at least nine feet. This meant that the Tangier mole must be built substantially higher than elsewhere in the Mediterranean – thus requiring an enormous amount of stone. Indeed, Sir Hugh observed that a plentiful supply of stone would be crucial for the undertaking.

One of Cholmley's first tasks when he arrived in Tangier in 1663 was therefore to find a good source of suitable stone, though he had the assurance that if such quarries were not found within three months, he could not be held to contract. The search was complicated by the steep cliffs on either side of the harbour, but in those to the west Cholmley was able to establish quarries, some half a mile from the Mole. But work on this structure was delayed by the need to build quarters for the men and horses, and workshops for the craftsmen. The accommodation was described by Sir Hugh as a 'square work', offering both shelter and protection. There was a 'handsome' courtyard in the middle, surrounded by buildings including the smith's shop, stables for 90 horses, cellars and magazines for all necessary stores, officers' quarters, and homes for those with families as well as lodgings for some 200 workmen and labourers. It was, wrote Sir Hugh, like a small city, named 'Whitby', after their home and his, by the original band of 40 craftsmen he had brought with him. These were boosted by other workmen and labourers as the work progressed, and to these were also added such soldiers and sailors as could be spared, although Sir Hugh was sometimes frustrated by having his workmen taken off the quarrying and mole-building to work on the fortifications. This contribution to the task of fortifying Tangier brought its own reward, however, for Whitby became one of the structures incorporated by the military engineer Sir Bernard de Gomme in his overall plan of defence against incursions by the Moors.[20]

Sir Hugh's initial scheme had been for the quarried stone to be carried by boat along the coast to the work on the Mole, but although the distance was according to him only 'about half a mile', the stormy seas and great tidal range presented a physical challenge, in addition to which the boats were often requisitioned for naval use. He therefore had wheeled carriages built and a coastal track constructed. This limited the loads that could be carried to four to five tons, although Sir Hugh came to recommend even lighter loads of three to four tons, for the convenience of handling.

★ ★ ★

The first matter of general interest arising from this study is the use of gunpowder in the quarrying of stone, and on this subject we are fortunate to have Sir Hugh's own description of the two methods employed, the smaller and the greater. The *smaller method* produced:

> lesser quantities by small mines, made after the common practice by drilling a hole of two inches bore nine or ten feet into the stone, conveying into this as much powder as is judged sufficient. The mouth is confined with plogs of wood, through which is bored a little hole, passing a cotton prepared with brimstone and other combustible matter, to convey fire to the powder, but because it is rare to bring down by such mines as these above 200 or 300 tons of stone, and that at Tangier there was occasion of far greater quantities, the practice there [and here Sir Hugh goes on to describe the second, *greater method*] is to have continually miners working in the main quarry such a mine as is usually made for blowing up the walls of fortified places, in which there is chiefly to be observed, that the turnings be made at right angles, and that the powder be lodged above the mouth or entrance of the mine, and the hole confined by filling up the mine. After the powder is placed with stone, and securing with good strong timbers the several angles that are made, I have known ten thousand tons of stone brought down in one of these mines with thirty barrels of powder, and it is remarkable, these greater mines, when they do most execution, make no noise, whereas the small drill mines, first mentioned, when sprung, give a report equal to a piece of cannon.

Sir Hugh observed that this stone was generally soft and therefore wasteful, but that the quarries at Tangier were 'capacious' enough to build several moles. He noted that he was to build one 400 yards in length and 200 in breadth [the arm at an angle already noted], and 25 to 30 yards in perpendicular height, and he concluded with the recommendation that 'whatever Prince shall hereafter undertake the like work, had need, before he begins, to be assured of like inexhaustible quarries'. A supply of gunpowder of the right quality and in sufficient quantity was also critical to the success of this work, especially in the face of rival military claims to that available, but this aspect of 'gunpowder engineering' requires further research and more information before this important part of a difficult problem can be resolved.

Sir Hugh's account of the lesser and the greater methods of gunpowder quarrying has a significance beyond that of this major project at Tangier. In terms of chronology the latter procedure came first, as a military response to the building of fortresses, with the first fully effective demonstrations coming at the siege of Sarzanello (Liguria) in 1487 and Castel Nuovo (Naples) in 1495. The procedures were discussed in a number of military manuals, but one of the best accounts is to be found in Vannoccio Biringuccio's *Pirotechnia* of 1540, written by a practical man of metals and fire.[21] He advised that the tunnel to the powder chamber should be long and winding (see Figure 2) so that the blast should not be carried back along it. The 30 barrels referred to in Sir Hugh's account are more than double the 14 shown in the illustration, although Biringuccio recommends that these should be supplemented by loose powder emptied around them. It is an irony of English history that many of the gentry (including the Cholmleys who backed first Parliament and then the King), would have been familiar with the creation of underground explosions from the years of the English Civil War, 1642-49. But whether those active at that time – the skilled miners undertaking this tunnelling and powder-setting, and the officers planning these operations – then applied the techniques to the winning of coal and quarrying of stone from the 1650s is not known, though Sir Hugh's account of producing large quantities of material for the Mole by this procedure suggests this experience.

Fig. 2. Vannoccio Biringuccio's sketch, showing the best plan for the twisting tunnel and barrels of powder in an underground mine.

It was probably however the lesser method described and employed by Sir Hugh at Tangier that held the key to the development of gunpowder quarrying and mining in England, because this was a less devastating procedure, producing stone or coal that was better formed and less friable. But little is known with certainty about when this was adopted for civil purposes and when it became widely used.[22] The subject was certainly under discussion in London in the early 1660s as the following extract from the Journal Book of the Royal Society for November 1663[23] indicates:

> Mr. Povey being called upon for Mr. Jonas Moore's way of breaking rocks with powder, Sir Robert related, from Prince Rupert, the following way of blowing up rocks underground in mines, viz by making a round hole in the rocks eight or ten inches deep, or as deep as they can, and then, by taking two wedges, which put together make a cylinder diagonally cut and by thrusting in the great end (after the powder is put in) and then driving in the other end: which done and fire being given to the powder by a train made in a groove in the wedges, the rocks are broken, because the wedges cannot be thrust out.

Although varying in some details, this account is very similar to Sir Hugh's description of his 'lesser method'. But what is also worthy of comment is the celebrity of the members of the Royal Society engaged in this discussion with their expert adviser Jonas Moore, and the enthusiasm with which they were discussing this practical matter. Those named include: Thomas Povey, Treasurer of the Tangier Committee of the Privy Council and uncle of William Blathwayt of Dyrham; Sir Robert Moray, the pre-charter President of the Royal Society; Prince Rupert, cousin of the King, member of the Tangier Committee, and involved in the company operating copper mines at Ecton, Staffordshire, where gunpowder is said to have been introduced in 1665; and Jonas Moore of the Board of Ordnance, surveyor and military engineer and recently back from his visit to Tangier with Sir Hugh Cholmley. This enthusiasm went on to include illustrations of the lesser shot-firing methods in the *Philosophical Transactions* of 1665.

It seems from this evidence, both practical, coming from Tangier, and theoretical, coming from the discussions of the Royal Society, that the greater and lesser gunpowder methods were being used in civil engineering from the early 1660s. But it also seems likely that, based on the experience gained during the English Civil Wars, blasting with gunpowder for civil purposes may have been introduced even earlier than that. This was probably the case at the Whitby quarries owned by the Cholmley family, from which stone was used in the harbour works there. Sir Hugh's final comment in the account quoted (on p.28) on the different noise level of the two methods indicates a familiarity with the respective procedures which suggests that he was not faced with a new situation in quarrying for the stone required at Tangier. He knew what he was talking about and had the experience to deal with the challenge posed by the construction of the Mole.

★ ★ ★

The second matter of general interest arising from Sir Hugh Cholmley's work on the Tangier Mole is the use of 'tarris' cement which set underwater. He wrote that the first foundations of the Mole were laid in August 1663, and that the stones were carried from the quarry in sailing boats some ten tons in burthen. As already noted, with the roughness of the tidal seas and other calls on the boats, waggons along a coastal track became the common form of transport. At the construction site the technique adopted, known as 'pierres perdues', was that of depositing a great mound of stones in the sea up to the low water level, along the chosen line of the Mole. This was then flat-topped with large stones cemented in position with lime and tarris, and clamped by bars of iron and lead to form a suitable bed for the roadway and buildings to be set up on the Mole. A drawing of 1670, prepared for the King when the Mole was 380 yards long, includes amongst the buildings listed a 'tarres house'. Unfortunately the features are blurred and this cannot be identified, but a later drawing of 1675 shows this to have been at a mid-point on the seaward side of the Mole.

In his 'Account of Tangier' Sir Hugh describes tarris as 'a certain sand made into mortar which hardens in the water', and goes on to observe that although 'Our Tangier tarras took some time to set; that obtained from Naples was very good'. This last reference takes us to the heart of the matter for tarris, so important in this engineering project because of its ability to set underwater, was the Roman or Pozzolana cement of ancient times. This was the powder from Putcoli (now Pozzuoli) on the north-west side of the Gulf of Naples. The ash was derived from volcanic eruptions, and it was discovered that mixed with lime these pozzolanic deposits produced an exceptional cement. The Pozzolana material was thus not itself a cement, but it contained silica and alumina which reacted with lime (produced by heating limestone) to form the 'hydraulic cement' that was to be so important for the underwater work at Tangier.[24] Its significance in this respect was well-known in earlier times. For example Vitruvius, who flourished in the later decades of the first century BC, wrote in his *De Architectura* of this powder that, 'being mixed with lime and rubble, [it] not only furnishes strength to other buildings, but also, when piers are built in the sea they set underwater'.[25] The surviving harbour works at Putcoli, themselves constructed with Pozzolana cement, demonstrated the significance of this strength. Sir Hugh's familiarity with these techniques described by Vitruvius may have come from his classical education, mentioned earlier.

By 1675 the Mole was 470 yards long. The drawing of that year prepared for the King has more detail and a greater clarity than that of 1670 referred to earlier. It is shown as Figure 3, overleaf. At the landward end of the Mole we see York Castle with the gate to Whitby (B) below this looming feature. Along the seaward side of the Mole there are storehouses, shops, lodgings and, of particular interest to this study, the Tarris Mill (N) and the 'Tarris beaters house' (q). 'S' marks the end of what is referred to in the key as 'the firm Tarris worke'. The task of construction continues beyond this for a short distance, and ends with a sunken Dutch man-of-war (T). The key tells us this was taken by Captain Harman, 'filled up with cemented matter and there sunk'. This ship had presumably been captured during the recent third Dutch War of 1672-4, a reminder of the military and naval significance of Tangier which is further emphasised by the line of guns along the seaward side of the Mole. In contrast, much of the harbour side of the Mole is taken up

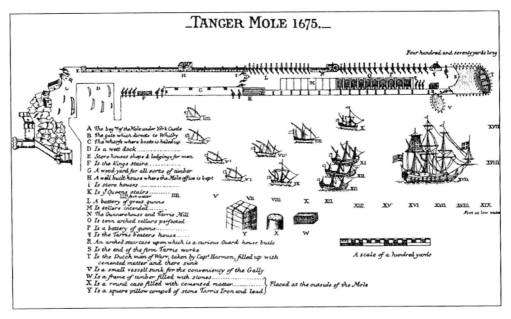

Fig.3. Plan of the Mole at Tangier 1675, showing the 'Tarris Mill' on the seaward side. (The advice of the Royal Archives, Windsor, is gratefully acknowledged)

with mercantile matters: the dry dock, the 'woodyard for all sorts of timber' (G), the 'Kings stairs' (F) and more elaborate 'Queens stairs' (H), and the fine range of warehouse-cellars, including 'tenn arched sellars perfected' (O) and eight in the process of fitting out (M).

Twelve ships are shown at anchor in the new harbour, mostly small vessels but with one large naval or merchant ship. More significant from our point of view, however, are the drawings of the stone columns that were stationed outside the Mole to provide protection against the breaking seas. The first type was the 'frame of timber filled with stones' (W); next the 'round case filled with cemented matter' (X); and lastly the most substantial of the three, the 'square pillow composd of stone Tarris Iron and lead'. In the drawing they are shown clustered round the end of the Mole, beyond 'the firm Tarris works' where this extra protection would be desirable. It should be mentioned that the previous drawing of 1670 included an illustration of pile driving which is not shown here. Wooden piles shod with iron had been an important earlier measure to protect the Mole against stormy seas. They were inserted amongst the debris of damaged sections by Sir Hugh in a system that he had successfully introduced at his home harbour of Whitby after he had observed the strength of a solitary pole placed as a marker for shipping, around which the waters swirled instead of crashing with destructive force. In Tangier however the piles were more susceptible to weakening by the activity of sea worms, and the introduction of stone columns set with tarris within a dispensable wooden frame must have offered a better solution to the problem.[26]

During the early and mid-1670s, Sir Hugh was engaged in active debate with his deputy Henry Sheres, artillery officer and military engineer.[27] Sheres was a strong advocate of the

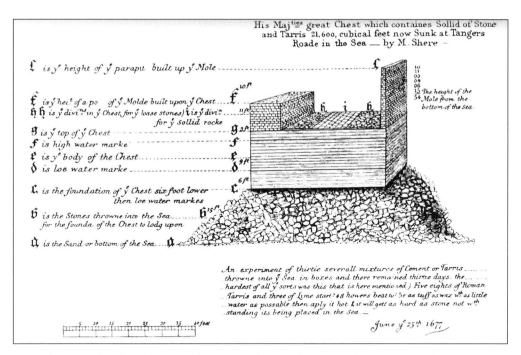

Fig. 4. The Great Chest, devised by Henry Sheres as part of a series of tests, 1677, from E.M.G. Routh, Tangier, *p.356. (The advice of the Royal Archives, Windsor, is gratefully acknowledged)*

'Chest' system of building the Mole rather than the 'pierres perdues' method favoured by Sir Hugh, whom he succeeded in 1676. Before going on to examine this debate further, the illustration of 'The Great Chest' in Figure 4 is unexpectedly worthy of extra scrutiny since it has as a bonus a written account of experiments carried out by Henry Sheres in the making of tarris. Dated 'June ye 25th 1677' it describes:

> An experiment of thirtie severall mixtures of Cement or Tarris throwne into ye Sea in boxes and there remained thirtie days. the hardest of all ye sorts was this that is here mentioned. Five eights of Roman Tarris and three of Lime startg + 8 houers beat to be as tuff as wax wth as little water as possable then aply it hot & it will gett as hard as stone not wth standing its being placed in the sea.

Henry Sheres's specific mention of the use of 'Roman Tarris' in his experiments, and Sir Hugh's observation mentioned earlier of the better quality of that from Naples compared to what was available in Tangier, indicates the use of Pozzolana cement in the construction of the Mole.

Lime may not always have been easily available, for in 1664 a request was sent to London for 'recruits, lime and workmen's tools'. The lime requested would have been quicklime, derived from limestone heated to a great temperature and then packed in sealed wooden barrels for a hazardous sea journey, during which water may have entered the barrels, thus dangerously 'slaking' the lime. It is not surprising that, given the quantities that would be required for the tarris making, one of Sir Bernard de Gomme's plans for that year refers to a new fort to be built near the lime kiln and the further end of the Mole.[28] The lime from the kiln would have been placed in a pit or box where it would be slaked with water before being mixed with the volcanic ash and sand. This was then subjected to a prolonged beating, probably of at least eight hours as suggested in the experiments described in Figure 4. It seems very likely that this was done in the 'Tarris mill' to which attention has already been drawn in Figure 3, and that it was undertaken by the 'Tarris beater' whose house was on the Mole, presumably because he had to live on the job. The drawing of the mill presents problems, because the wheel seems to represent a water wheel, which was very unlikely given its position well above sea level on the roadway of the Mole. It is more likely that the power came from a windmill of the Mediterranean rather than the northern European style, one that was made up of numerous spokes, usually without sails, set within a rim to give stability.[29] This device may have powered a set of stamps as used in the early gunpowder mills when it was realised that the three chief ingredients must be moistened and incorporated under pressure rather than just mixed together. Similarly in the case of tarris, the ingredients had to be mixed so intimately that they became bound together like wax – impenetrable to water. In both cases, this took hours of beating.

It is an intriguing aspect of this subject that although the Roman cement of the ancient world continued to be used in the Mediterranean region, as shown by its importance in seventeenth-century Tangier, the knowledge of this useful technique seems to have faded in western Europe. Bélidor undertook research in France, publishing on the subject in 1729, but in 1756 John Smeaton had to find by experiment the best sources for the ingredients of a cement able to harden rapidly in the offshore waters where the Eddystone Lighthouse was to be rebuilt. He chose hydraulic lime from Aberthaw and pozzolana from Civita Vecchia.[30]

★ ★ ★

'The Great Chest Constructed by Mr Sheres, June 1677' may also be interpreted as exemplifying a further and more general aspect of this work of engineering, and that is the awareness of some English engineers of similar work undertaken in Mediterranean ports. Sir Hugh's familiarity with the harbour works, anciently at Ostia and more recently at Genoa, has already been commented upon, and it may also be noted that in 1663 he sought the advice of Genoese engineers by arranging a visit to Tangier.[32] He decided against adopting the chest method but in a letter to the Lords Commissioners in 1665 he revealed that he was then nevertheless employing men from such harbours. Their experience might be thought a good enough reason, but Sir Hugh's letter shows that their sobriety was also valued. He wrote:

> My Lords I have found some Itallians and others much freer from drink than our men are which hath caused me to send for some of these Ordinary workmen and some others that have bin experienced in the work of the Mole at Genoway, I am taking care also to do the same at Marsells and question not but to be well supplied from these parts, if the fear of the Turk does not frighten them from taking their Jorny hither …[32]

But despite his willingness to draw on the labour of the ordinary and sober workmen and the experience of the 'Genoway' engineers, and to investigate the matter further in Marseilles, Sir Hugh remained sceptical about the relevance to Tangier of the 'great chest' method. He argued that the great tidal range and unpredictability of wind and waves at Tangier would make it very difficult to lower and secure the unwieldy chests on the rocky base formed on the seabed, with the precision evident in the calmer waters at Genoa where divers could be employed to level the foundations. But he was hard-pressed on the matter by Henry Sheres who had gained experience at Tangier by taking charge when Sir Hugh was called to London on family matters or on problems with regard to delays in financing the project. At the re-organisation in 1669 following the death of the other two contractors, Sir Hugh was appointed Surveyor General to the Office of the Mole with Sheres as his Clerk of Works. Despite this confirmation of Sir Hugh's position, his preferred method of working and so also his authority was being weakened by the damage caused by rough seas breaking in upon the Mole. Henry Sheres travelled to Genoa to see the situation there for himself, and he returned convinced that this was the way ahead. Sir Hugh agreed to allow the first experiment with the chest method to be undertaken in 1670, but as he had anticipated matters did not go well. In the course of several attempts to lay the chest Sheres was injured and had to call out Sir Hugh to rescue the operation in the face of the strong Levant wind that was blowing. This did not settle the matter however for despite the problems shown in Figure 4, especially the unwieldy nature of a large and weighty wooden cage that must settle on a bed of 'Stones throwne into the Sea', this way of building up the Mole gradually gained supporters in London. Sir Hugh showed flexibility by agreeing to a modified version of the Genoese method, but with his authority further challenged by an agreement to complete the work on Sheres's plan rather than his, he retired in 1676 and Henry Sheres was appointed in his place.[33]

The features on the superstructure had already become well-established in Sir Hugh's time, and this pattern continued to be observed as the Mole reached its final probe out to sea. They included a slender parapet on the seaward side and a more stocky wall on the harbour side, capable of accommodating the cellar-warehouses needed for the mercantile activities of the port. The batteries of guns shown in Figure 3 provide a reminder of the military and naval significance of the Mole, which must be sturdy enough to withstand their reverberations when fired. Movement along the surface of the Mole was facilitated by a paved central roadway some 20 feet wide, edged with 'pavements' of loose stones on either side. Nearing completion when Sir Hugh left, the Mole was much admired by visitors as may be seen from the following comments made in the year of his departure, 1676, even allowing for some hyperbole:

> The Mole is in its design the greatest and most noble Undertaking in the World, it is a very pleasant thing to look on … now near 470 yards long, and 30 yards broad, several pretty Houses upon it and many Families; on the inner side 24 Arched Cellars and before them a curious Walk, with Pillars for the Mooring of Ships. Upon the Mole are a vast number of Great Guns, wch are almost continually kept warm during fine weather, in giving and paying Salutes to ships which come in and out.[34]

Despite the success of the Mole in aesthetic and engineering terms, and its usefulness as a safe harbour for the careening of naval ships guarding the seas in the interests of both the state and commerce, its future was placed in doubt by the unwillingness of either the King (for whom this was after all a personal possession) or Parliament (which was increasingly unwilling to finance his extravagances), to provide the funds necessary to maintain this settlement. In addition, there was a lessening of the threat of piracy, and a rising challenge from the Moors which culminated in the great siege of 1680 and the withdrawal of the English three to four years later. It is at this point that we see the last of the neglected themes mentioned earlier: namely the relationship between the growing profession of the military engineer such as Henry Sheres, and the continuing independence of the contracting civil engineer such as Sir Hugh Cholmley. The great rivalry between the two over the engineering challenge posed by the Mole personifies the contrast at that time between the mathematically-skilled artillery officer, secure within a Board of Ordnance system reorganised and strengthened after the divisions of the civil wars, familiar with problems of construction and fortification, and the entrepreneurial gentleman, a member of a class not a profession, experienced in developing his economic interests at home and overseas, even though this involved in Sir Hugh's case the pursuit of the sale of alum from his family mines to merchants in Leghorn, whilst deeply involved with his responsibilities in Tangier.[35] On retirement he could return to his estate and economic interests in Whitby, taking up the responsibilities of a country gentleman with an inherited title, representing the interests of his constituency and himself in Parliament. Henry Sheres on the other hand must shift his attention not to the relative calm of the English countryside but to the defence of Tangier against the Moors, and a continuing military commitment to the artillery that would lead him shortly to an unwelcome role in the Sedgemoor campaign of 1685. Here as he reported to Lord Dartmouth, the Master General of the Board of Ordnance, he saw 'too much violence and wickedness practised to be fond of this trade … for what we every day practise among this poor people cannot be supported by any man of the least morality'. He was however to be rewarded for his services with a knighthood, and appointment to the post of Surveyor of the Ordnance which brought responsibility for the engineers employed, many of whom would be civilians performing particular jobs.[36]

In the meantime the concluding drama at Tangier may be illustrated by reference to Figure 5 where the Mole is shown as an empty abandoned concept, the focus now being entirely on the land defences, brought to a state of preparedness in anticipation of the advance of the encircling Moors. The plan shows the city wall, protected by rings of outlying forts of which those numbered 7, 8, 9, 10 and 11 and named 'Forlorn Hope', must have daunted even the bravest hearts. Also marked is the small fort and village of 'Whitby' where Sir Hugh's original workforce lived, within the outer rim of fortifications. In March 1679 Whitby was the specific target of a fierce attack on the western lines launched by a now well-trained and substantial army under the command of the Alcaïd Omar of Alcazar. In a last desperate effort the sergeant, having ordered his surviving

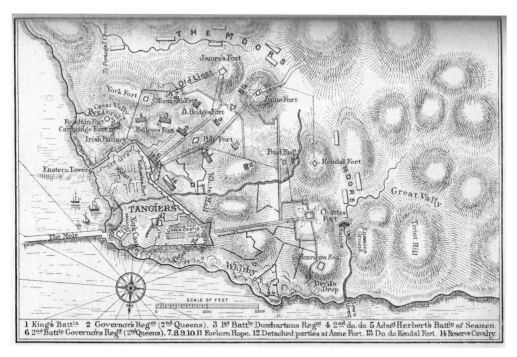

Fig. 5. The defences at Tangier, 1680, from E.M.G. Routh, Tangier, *p. 194. Note that Whitby is included in the battle plan.*

men to retreat, blew up the store of powder, killing himself and 40 of the attackers.[37] Perhaps this was the powder for the quarries and it may be that the Moors were hoping to make good their own chronic shortage of powder from this magazine. A nominal truce was established in the summer but the powerful Emperor in Fez, Mulaï Ismâïl, refused to confirm this and fighting was resumed the following year as the Moors sought to expel the English and take over the harbour and Mole whose construction had been closely observed by the watching army in the hills, shown so clearly in Figure 1, their positions shown in Figure 5. Unfortunately for the defenders of this base, especially those convinced of its strategic importance, it was already being decided in London that Tangier should not continue to be held – it must be abandoned.

We know from contemporary drawings and the accounts of travellers that the completed work with its many facilities was a major achievement of its time. But in order to justify its demolition and that of the town, this achievement had to be denigrated. By a great irony the military engineer charged with writing the critical report and then supervising the demolition of the Mole was Henry Sheres. We have a good account of this operation by an eye-witness – Samuel Pepys – chosen by Lord Dartmouth to accompany him on his mission of withdrawal and destruction in order to evaluate the compensation to be paid to those losing property, businesses, and livelihood. After some earlier friction in London between Pepys and Sheres, due to the former thinking that the latter was paying undue attention to his wife, the diarist had come to respect the professional expertise of the military engineer. Evidence of the strong degree of confidence that came to exist between them is shown by Pepys's diary entry for 30 October 1683, that Sheres 'did show me his foul draft of the ordinary objections made against the Mole [,] improved the most he could to justify the King's destroying of it, though he did tell me privately that he is able to answer them all'.[38]

Pepys's account of the difficulties experienced in blowing up the Mole, caused by the strength of the structure especially the arched warehouses that were only brought down after much effort, is a tribute to the widespread use of tarris. He observed the methods used, including the iron cylinders which may have facilitated the insertion of powder charges into the stonework. He recorded in his diary entry for 24 October 1663, for example, that having observed the explosion of a mine at a distance, he went on to the Mole where he saw 'some parcels of the iron cylinder making their way quite through the side of the Mole and making a crack in one place quite across the Mole from side to side, and yet there was not a full barrel and a half of powder [used]'. Lord Dartmouth was well aware of what a time-consuming work this was proving to be, especially the demolition of the 'body of Sir Hugh Cholmlye's parte wch is very great and must be removed into the harbour by hands and labour' after the initial loosening by powder.[39] But even with 2,000 soldiers and sailors at work until late each night, the task was not completed and the Mole levelled until January 1684, when the last ships left the now rock-strewn and useless harbour. On board was Lord Dartmouth, fearful according to Pepys in his diary entry of 29th March 1684, that the way was now open to the French to assert their ascendancy.

★ ★ ★

And so we return to the painting at Dyrham, showing the evacuation of Tangier with the Moorish soldiers waiting to take possession of the work created by civil and military engineers, whilst the English destroy the town, harbour, and Mole as they retreat towards the sea. The painting echoes the vision of the almost spectral figures observed by Pepys on the fine evening of 24 October 1683 as he trained his 'long glass' on them and 'saw the whole camp of the Moors and their huts and manner of walking up and down in their alhaques [so] that they look almost like ghosts all in white'. How did this powerful painting come to Dyrham? Was it through the intervention of Thomas Povey, William Blathwayt's uncle, the former Treasurer of the Tangier Committee and noted patron of the arts? Is it here because this depiction of failure was thought more suitable for the walls of a country house than for those of a mansion or possibly even a palace in the metropolis? Or more prosaically did William Blathwayt, as the executor of his uncle's will, bring it to the country house he was re-building, as a usefully large painting for an empty wall? And in any case, should the Tangier episode be regarded as a failure, or as an opportunity from which lessons were to be learnt that would lead to the establishment of Britain as a great engineering, trading and military power?

First of all a great technological advance should be recognised, for it may be argued that the successful creation of the Mole represented 'The Birth of British Gunpowder Engineering Overseas', especially as this project was followed by many others beyond

Europe that relied on explosives to clear the way for the roads and railways, irrigation and harbour works that were to be built by British engineers. A similar claim for the use of tarris or hydraulic cement, especially for underwater work, is more of a problem since although this was vital for the building of the Mole the technique seems then to have been lost for many years, requiring rediscovery in the mid-eighteenth century. Next, several institutions were created or revived by this experience, especially the Army and the Navy. Several regiments proudly trace their origin to service in Tangier, and although the loss of this harbour was a particular blow to the navy, the lesson had been learnt and in a short time a permanent base was to be found in Gibraltar. The Board of Ordnance, which had suffered greatly from the political divisions of the previous decades, was strengthened by this demonstration of the professionalism and broad talents of its military engineers. Even the Royal Society, its charter granted by Charles II after his restoration, gained from the discussion of the challenges posed by technological problems. But the construction of this fine work of engineering did nothing to further the recognition of the profession of civil engineering even though the Mole was a work of two parts, the responsibility of Sir Hugh Cholmley from 1663 to 1676, and then of Henry Sheres from 1676 to 1683. Both should surely be recognised as engineers, whether civil or military, but the *Biographical Dictionary of Civil Engineers* defines the former as 'contractor' and the latter as 'military engineer'.

Thirdly the history of Tangier showed the importance of a settled state at home and secure lines of supply. Governors, engineers and soldiers were kept chronically short of supplies throughout the period of this experiment. In part this was because this was not a State project. Tangier was a personal possession of the Crown, and when Charles appealed to Parliament for financial help this was seen as an opportunity to enforce agreement in important matters such as the succession to the throne. The absence of legitimate children made Charles vulnerable in this respect for his strongly Catholic brother and heir, James, was not welcome. But in this Exclusion Crisis Charles II refused to exclude his brother, so Parliamentary funds were not forthcoming, and Tangier had to go, especially as public support for the enterprise was waning. Unsettled circumstances at home could thus prejudice overseas interests.

In certain circumstances however, the establishment of a strong mercantile community could overcome this problem, a thought prompted by the contrasting fortunes of Tangier and Bombay, both part of Catherine's dowry. Bombay on the west coast of India was considered to be so far away as to be of lesser importance. It was therefore handed by Charles to the East India Company (established in 1600) for a peppercorn rent. Serving self interest the Company flourished from the trade it was able to conduct through agreements with the Indian princes who, unlike these Moorish leaders, provided a structure of stability within India. In contrast, the whole relationship with the Moors seems to have suffered from the lack of any guiding policy issued from London or Tangier. It does seem remarkable that at a time when the Tangier enclave was particularly threatened in the early 1680s, Charles II should have been placing an order with the Board of Ordnance for '221 Barbary gun barrels with locks and cases …' to be delivered into His Majesty's stores 'for a present to the Emperor of Morocco and Fez', perhaps in the misguided hope of buying peace through diplomacy.[40] More practical and more sinister was the pursuit by the Moorish military leaders of war materials, especially gunpowder, from English officials and merchants. It has been claimed that Morocco underwent its own military revolution in the sixteenth century but there is little evidence of that in these Tangier years when its resources for gunpowder conflict were limited.[41] Catharine's dowry had presented opportunities for steps forward in technology and trade, but there remained much to be learnt about government and diplomatic skills if powerful ambitions in the Mediterranean and beyond were to be realised, and old enemies such as France, held in check.

Notes

1. For information on the family and the house see the well-illustrated National Trust booklet, *Dyrham Park* (2000), text by Oliver Garnett.
2. Thomas Povey (1614-c.1705) is described in the *Oxford Dictionary of National Biography* (Oxford: Oxford University Press, 2005), referred to hereafter as ODNB, as a colonial entrepreneur and administrator who gained greatly from his many government posts. These included that of Treasurer for Tangier from October 1662 until Samuel Pepys succeeded him in 1665. Regarded as a man of taste he spent much on paintings and books. After his death his nephew became the administrator of his estate in July 1705. Like the uncle who brought him up, William Blathwayt (bap.1650-1717) gained profit and power from government service, but with a better reputation for competence than Povey.
3. *The Handlist of British Diplomatic Representatives 1509-1688* by Gary M. Bell (Cambridge: Cambridge University Press for the Royal Historical Society, 1990), shows that Blathwayt served in the Low Countries July 1668-October 1670 and June 1671-January 1672; in Sweden February 1670-September 1672; and Denmark February 1672-December 1672.
4. These and other goods were largely the 'fruits of office' as is shown by Barbara Murison in her article 'Getting and Spending. William Blathwayt and Dyrham Park' in *History Today* (1990), vol.40, pp.22-28, and her biography of Blathwayt in the *ODNB*.
5. The 1912 account by E.M.G. Routh entitled *Tangier: England's Lost Atlantic Outpost, 1661-1684* (London: John Murray, 1912), remains the best scholarly book-length study of the subject.
6. See *Captives* (London: Jonathan Cape, 2002) by Linda Colley for an account of this challenge, and also for a chapter on Tangier.
7. A.J. Smithers, *The Tangier Campaign: The Birth of the British Army* (Stroud: Tempus, 2003), deals with this aspect of the subject well, but without identifying his sources. The Dyrham painting on the cover is shown in reverse, thus losing the point of the protection provided by the Mole against storms from the west. See also *The History of the Second Queen's Royal Regiment, now the Queen's (Royal West Surrey) Regiment, Vol. 1: The English occupation of Tangiers, from 1661 to 1684* by Lt-Col. John Davis (London: Bentley, 1887), for an account of the putting together of an army from disparate sources. I am grateful to Mr W. Hanna for drawing my attention to this volume.
8. Royal Proclamations in the library of the Society of Antiquaries of London reveal the King's ambitions. From 29 September 1662 a Free Port was to operate so that merchants 'shall have a good port in the entry to the Mediterranean to befriend them'; and an order of 13 January 1674/5 assured merchants [in the uncertain situation of the third Dutch War ending but French hostilities continuing] that should war be declared between the King and their own Prince, goods and effects would not be seized for a further six months.
9. Routh, *Tangier*, pp.19-22.
10. Smithers, *Tangier Campaign*, pp.34-5. In March 1662 Gayland's price for an armistice of six months was 50 barrels of powder.
11. Edward Montagu (1625-72), the first Earl of Sandwich, supported the Protectorate but went over to the Royalists in 1659. He brought home Charles II from exile in his flagship the *Naseby*, renamed the *Royal Charles*. C.P. Hill, *Who's Who in Stuart Britain* (London: Shepheard Walwyn, 1988), p.239.

12 Martin Beckman (1635-1702), knighted 1685. Appointed Engineer to the Ordnance in 1770 and promoted to Chief Engineer in 1685. Beckman's long association with Tangier spanned the first survey with Lord Sandwich (1661-2); his subsequent appointment as Engineer General there with responsibility (with Bernard de Gomme) for the fortifications; and his final and sole responsibility for the demolition of the fortress on the expedition led by Lord Dartmouth (1683-4); *ODNB*; and the *Biographical Dictionary of Civil Engineers in Great Britain and Ireland* (London: The Institution of Civil Engineers, 2002), vol.1: 1500-1830, edited by A.W. Skempton et al.

13 Routh, *Tangier*, pp.30-34, 343-4 and Appendix IV, Accounts for the Mole. Also Frances Willmoth, *Sir Jonas Moore: Practical Mathematics and Restoration Science* (Woodbridge: Boydell Press, 1993), p.131.

14 Hugh Cholmley (1632-89), 4th Baronet 1665, has no entry in the *DNB* or *ODNB*, but receives recognition in the *Dictionary of Civil Engineers*. His reputation was based on his work at Whitby harbour where his family had strong landed and economic interests. One of the three original contractors, the arrangements were revised in August 1669 following the death of his partners, and he was appointed Surveyor General of the Office of the Mole. He retired in 1676 and the position was taken by Henry Sheres.

15 *The Diary of Samuel Pepys*, edited by Robert Latham and William Matthews (London: Harper Collins, 2000), vol.VIII, 1667, pp.592-3, 27 December 1667. Pepys had asked Sir Hugh about the state of his business accounts should he suddenly die. He was told that all could be easily understood except 'a sum of 500£ which he hath entered "Given to E.E.S.", which in great confidence he did discover to me to be my Lord Sandwich, at the beginning of their contract for the Molle'. Pepys supposes that the other two contractors would have given the same sum.

16 G. Young, *A History of Whitby* (Whitby: Clark & Medd, 1817), 2 vols.; *Victoria County History, County of the North Riding, Yorkshire* (London: Constable, 1923), vol.2; G.H.J. Daysh, ed., *A Survey of Whitby and the Surrounding Area* (Eton: Shakespeare Head Press, 1958); Andrew White, *A History of Whitby* (Chichester: Phillimore, 2004).

17 Jonas Moore (1617-79,) knighted 1673. See the *ODNB* entry by his biographer Willmoth (n.13), also the entry in *Dictionary of Civil Engineers*. This mathematician, surveyor and astronomer made his name with a 16 sheet 'Mapp of the Great Levell' of the fens (1657-8), following this with surveys of Tangier and the 'Bounds of the Mole' in 1664. He became Assistant Surveyor to the Board of Ordnance in 1665 and Surveyor General in 1669. He was one of last of the Ordnance officers to go on active service – in the 3rd Dutch War.

18 For example the drawings and engravings made by Wenceslaus Hollar during his visit of September 1669.

19 See 'An Account of Tangier by Sir Hugh Cholmley Bt., with some account of himself and his journey through France and Spain to that Place where he was engaged in building the Mole in the Time of King Charles the Second; and a Journal of the Work carrying on', assembled by his kinsman Nathaniel Cholmley of Whitby and Howsham Esq. from manuscripts in his possession, 1787, in the Bodleian Library, Oxford. There are also papers and letter books in the North Yorkshire Record Office (NYRO). I am grateful to this Office for forwarding information to me.

20 Sir Bernard de Gomme (1620-85), knighted by Charles I. Dutch by birth he came to England with Prince Rupert and became Engineer and Quartermaster General to the King's forces during the Civil War. After the Restoration he became Chief Engineer to the Ordnance (1661-85) and Surveyor General (1682-5), *ODNB*. Andrew Saunders' account of the *Fortress Builder Bernard de Gomme, Charles II's Military Engineer* (Exeter: University of Exeter Press, 2004), p.342, plan 81, refers to the 'Description of his Maties [Majesty's] Buildings at Whitby belonging to ye affaires of ye Molle at Tangier', National Archives.

21 *The Pirotechnia of Vannoccio Biringuccio* (1540), trans. from the Italian with intro. and notes by Cyril Stanley Smith and Martha Teach Gnudi (Cambridge, Mass.: MIT Press, 1942, reissued 1959), pp.422-5.

22 I am grateful to Professor Raffaello Vergani for sending me an English version of his paper on 'The civil uses of gunpowder (15th-18th centuries): a reappraisal', published in Italian in 2003. He finds the earliest use of powder in stone quarrying in 1621 in Saxony (where the weight of the charge was 11-12 pounds). But the experiment was costly and perhaps not widely followed up. The next case mentioned is in England, in a report of 1671 by Robert Boyle, quoted in 'Gunpowder and Mining in Sixteenth- and Seventeenth-Century Europe', *History of Technology* (1985), vol.10, by G.J. Hollister-Short, who refers to the traditional timing for the introduction of blasting at copper mines (not stone quarries) as 1670, perhaps between 1665 and 1680.

23 Thomas Birch, *History of the Royal Society of London* (London: A. Millar, 1756), vol.1, p.335, 16 November 1663.

24 P.E. Halstead, 'The Early History of Portland Cement', *Transactions of the Newcomen Society* (1961-62), vol.XXXIV, pp.37-54; P.C. Kotzias, 'Volcanic Ash in Ancient and Modern Construction', The Thera Foundation, 3rd International Congress, Santorini, Greece, 1989.

25 Vitruvius, *Ten Books on Architecture*, translated by Ingrid D. Rowland, edited by Rowland and Thomas Noble Howe (Cambridge: Cambridge University Press, 1999), pp.36-38, 73-74, 179-180; Mike Jackson *et al*, 'Pozzolana Cement', *The IRM Quarterly* (2003), vol.13, no.3.

26 In a letter of 1665 to the Lords Commissioners, *The Cholmleys of Whitby* (NYRO), pp.7-8, Sir Hugh observed that 'the worm does so abound in these seas' rotting wooden foundations. Routh, *Tangier*, p.351, notes that in 1677 Pepys sent to Sheres 'a small case of temper'd stuff designed for the killing of the worme', which the King would like to be used for experiments in the Mole.

27 Henry Sheres (bap.1641-1710), knighted 1685, FRS 1675. Military engineer long associated with Tangier, undertaking surveys in 1668, becoming Surveyor General of the Mole in 1676, completing its construction but then superintending its demolition in 1683-4. *ODNB*, *Dictionary of Civil Engineers*.

28 John P. Speweik, 'Lime's role in mortar', *Masonry Construction* (Aberdeen Group), Vol.9, August 1996, pp.364-368; Routh, *Tangier*, p.72; Saunders, *Fortress Builder*, pp.94, 96.

29 I am grateful to Owen Ward for his advice on the Mediterranean-style windmills, and for drawing to my attention the well-illustrated book by Alejandro García Llinás, *Die Windmuehlen der Balearen* (Palma de Mallorca), 1995.

30 Halstead, 'Early History of Portland Cement', pp.41-5. It is noted in the *Dictionary of Civil Engineers*, pp.621-2, that Smeaton's experimental work on Pozzolana cement *only* [my italics] became known in the 1790s after the publication of his *Narrative of the Building … of the Edystone Lighthouse* (London: G. Nicol, 1791).

31 *Dictionary of Civil Engineers*, p.134. In the entry for Sheres, pp.603-4, it is noted that in 1664 Signor Jacomo, a Genoan, had offered to undertake work by the chest method, but this was rejected.

32 Letter from Sir Hugh Cholmley to the Lords Commissioners, 1765, in *The Cholmleys of Whitby*, p.7.

33 For a good account of these manoeuvres see Routh, *Tangier*, chapter 17.

34 Routh, *Tangier*, pp.354-6.

35 Letter from Sir Hugh Cholmley in Plymouth, about to leave for Tangier in late 1664, to an agent in Leghorn concerning 'the accompt of the former allome', *The Cholmleys of Whitby*, p.9.

36 Lord Dartmouth (1648-91) was formerly George Legge, a naval officer created the first baron in 1682. Like his father William he had a long association with the Ordnance, becoming Lieutenant General in 1672 and Master General in 1682. In 1683 he received the secret commission to organise the abandonment of Tangier. *ODNB*; D.M.O. Miller, *The Master-General of the Ordnance* (n.d.1972?), pp.44-5; *The Diary of Samuel Pepys: Companion*, pp.412-13. For Sheres's letter see the entry under his name in *ODNB*.

37 A bounty of £60 was paid to the sergeant's widow, her husband having blown up 'Whitley [Whitby] Fort at Tangier, and lost his life there', 25 April 1680, Routh, *Tangier*, p.166.

38 See Samuel Pepys's 'Tangier Journal, 30 July-1 December 1683' in *Pepys's Later Diaries*, ed. C.S. Knighton (Stroud: Sutton Publishing, 2004), 30 October 1683; *ODNB*; Routh, *Tangier*, p.359, n.1.

39 Routh, *Tangier*, pp.362-3.

40 John Cooper, 'Thomas Hawley, Gunmaker', *Journal of the Arms and Armour Society* (vol.xix, 4, 2008), pp.164-176.

41 Weston F. Cook, Jr., *The Hundred Years War for Morocco: Gunpowder and the Military Revolution in the Early Modern Muslim World* (Boulder, Co.,: Westview Press, 1994).

3: The Lost Distilleries of Bristol and Bath 1775–1815

Mike Bone

In his many books on engineers, engineering and transport history L.T.C. Rolt made few references to the production of food and drink and, with the exception of a proposed volume on farm machinery, the area did not feature in the Longman's Industrial Archaeology Series which he edited from 1969. One might, therefore, question the inclusion of an article on distilling in two of England's provincial cities in a volume to commemorate his birth and to explore the themes of landscape and technology. The industry in England has been somewhat neglected by historians but its innovative role in making use of the steam power and engineering that L.T.C. Rolt was largely concerned with provides a fascinating case study of technological advance. It also had a significant impact on the urban landscape and the lives of the people, thus touching on the social consequences of technological development, a theme that he was to return to at the end of one of his later books and remains an important issue to this day (Rolt 1970).

The Development of a Distilling Industry

Distillation produces a strong spirit by heating and then condensing a fermented liquid to separate alcohol and water, the respective boiling points of these being 180°F and 212°F. Its early history in the West is uncertain: it seems that it arrived in Europe from the Middle East by 1100 AD, the words 'alcohol' and 'alembic' – an old word for a still – being Arabic in origin. Early uses of distilled liquors were medicinal and physicians were probably the first to distil in England. Leonardo, who designed improved alembics, recommended its addition to pigments and gunpowder but warned against drinking spirits. However, the line between medicinal use and pleasure has often been a fine one and use as a beverage in England dates from Tudor times and the rise of new consumption industries noted by Joan Thirsk (1978, pp.93-96).

Distillation by physicians was to continue and country houses would often produce their own remedies and cordials – a Still House with 'a Pewter Lembick & Copper pot' and 'a cold Still' are included in an inventory of 1710 from Dyrham House, near Bath. (Walton 1986, p.68). Distilling as a discrete profession dates from the reign of James I when the expansion of trade and England's maritime enterprise stimulated demand for beverages that kept better than other liquids and, suitably diluted, could be used by ships crews and for foreign trade. The early industry was concentrated in London, Liverpool, Plymouth and Bristol but a poor product – the Scots and Irish made spirits that tasted far better – restricted output, as did competition from imported brandy (distilled from wine) and rum made from by-products of a growing sugar trade (Harper 1999 and Berlin 1996).

The Dutch connection was to change things – they had distilled basic spirit (*aqua vitae*) with oil of juniper to produce a new beverage which became known as 'genièvre' in France and 'geneva' or 'gin' in England. English distillers were slow to realise the potential of gin but war with France in 1688 and the accession of William III, led to the prohibition of brandy imports and tax incentives to encourage the use of English grain surpluses in the stills. Quality was also improved and there was a massive increase of 980 per cent in output of spirits between 1684 and 1754 (Coleman 1977, p.120). Distilling, therefore, became wedded to England's prosperity in agriculture and overseas trade. The industry also contributed to food production by using its considerable waste products of spent grains and liquids (wash) for stock-rearing (Mathias 1979, pp.252-264). The distilleries also stimulated demand for coal and metal goods in the operation of large and integrated plants that were also to pioneer new forms of business organisation.

The downside of the rise of the distillery was an age of drunkenness that has dominated historical writing on the industry and its products. Excessive gin drinking, encouraged by low taxation and a proliferation of sales outlets, was a particular problem in London and some of England's larger cities. Opposition to cheap spirits came to a head in the mid-eighteenth century when the author and magistrate Henry Fielding, the artist William Hogarth – his 'Gin Lane' was a particularly savage indictment of binge drinking – and others such as Joseph Tucker, rector of All Saints in Bristol, made a powerful case for tighter control of manufacture and sale of gin which led to the first effective 'Gin Act' of 1751. Thereafter, restrictions on production by the Excise, control of sales outlets and the end of grain surpluses as the population grew, led to a decline in output and the number of distilleries. The distilleries were closed by government during times of bad harvests and fear of famine and malt distillers also had to contend with the influence of the West Indian planters who sought outlets for surplus molasses and sugar (Harper 1999, Dillon 2002 and Warner 2003).

In the Distillery: Process and Equipment

Distillation is a fairly simple process and can be carried out on a small scale, as it has been by those eager to avoid excise duties on spirits over the years. In a commercial context, all kinds of grain, raw or malted, plus juices of fruits, sugar cane, root crops and other vegetables can be converted into alcohol prior to distillation. Preparation varies with the ingredient – French brandy is made from wine, and West Indian rum from sugar/molasses dissolved in water – but British spirit is made primarily from corn and can then be subsequently re-distilled to produce whisky, gin, British Brandy or British Rum and other compounds.

Corn was prepared for distillation by a process similar to brewing. Malted grain was first crushed between rollers in a mill and raw grain was coarsely milled between stones. The grist was then mixed with hot water (or mashed) in a large tun to convert the starches in the grain into sugars. The resulting liquid was run off into an underback and then cooled in large, shallow vessels prior to fermentation in backs using brewers' yeast. The fermented wash was then passed to a charger, a large iron tank, and then fed into a wash still. When the alcohol had evaporated, the spent wash was taken off to add to animal feed as an aid to fattening of pigs and cattle. The vapour from the still was condensed in a worm immersed in a tub of cold water and this spirit ('low wines') was passed to a worm safe to test its strength. It then flowed to a low-wines receiver and was redistilled in a low wines or spirit still. During this process impurities called 'feints' were removed and redistilled. The spirit would be stored in a spirit safe and store vats and was known as plain British Spirits.

This spirit then passed to the rectifier for conversion to whisky, gin, brandy, etc. Here the spirit was first redistilled to rectify or purify it, and then again, with herbs, berries, or other 'botanicals' added to impart flavour. Rectifying stills were smaller than those in primary distilleries and separate stills were often kept for different products. The product was then stored. The whole process was closely controlled by the excise officers as, before regular income tax, the government relied heavily on taxation of malt and the drink trades (Dodd 1843, pp.41-62 and Bateman 1840, pp.340-341).

The Industry in Bath and Bristol

Whilst there are scattered references to the making and consumption of spirits in Bristol and Bath in earlier centuries, it is only possible to get a clear picture of the extent and nature of the trade when local and national directories and newspapers appear from the 1770s. Gin was certainly consumed in both cities as evidenced by the mock funerals for 'Mother Gin' held by way of popular protest following an Act to restrict the supply of the drink in 1736. The riots feared by government did not materialise, the occasions being marked instead by widespread drunkenness (Warner 2003, pp.126-9; Latimer 1893, p.198). Latimer notes the erection of a number of distilleries in Bristol at the outbreak of war in 1689, encouraged by the stopping of French brandy imports and the cheapness of coal (Latimer 1893, p.7) and there were protests by the distillers in 1713 after restoration of trade with France threatened their businesses (Latimer 1893, p.101) A few probate inventories of distillers also survive for the city from this early period (George and George, 2005 pp.115-6 and 2008, pp.129-30). A revival in the trade after the slump which followed the 1751 Gin Act came after duties were reduced in 1785 and annual output of British spirits increased from an average 2.5 to 4.5 million gallons until the end of the century (Ashton 1972, pp.57, 243). This prosperity is reflected in contemporary local publications, Barrett noting in 1789 that 'The distillery is also become a very capital branch of trade, many great works being erected at amazing expense in different parts of the city ...'. He also noted the scarcity of grain for food, as the distillers drew supplies from Gloucestershire, Wiltshire, Worcestershire, Herefordshire and Wales, and also the great export of spirits to North America, Africa and London – the latter with 'such large distilleries ... as exceeds all belief', a reference reported in Heath's later guidebook (Barrett 1789, p.185, Heath 1799, p.70).

The first street directory for Bath dates from 1773 and has few commercial entries and no distillers. The only references in our period are to Bourne's distillery on the Quay in the city in the *New Bath Directory* in 1805 and 1809. This contrasts markedly with the situation in Bristol, as summarised in Table 1. This attempts to show the number of distillers – the difference between primary or malt distillers and rectifiers is not consistently made – from Sketchley's directory of 1775, other national and provincial directories and, finally, from the superb run of Matthews' (later Mathews) directories which began in 1793. In addition to the total number of directory entries, an estimate of the total firms presumed active in any one year has been made by adding those listed in previous or subsequent years, on the assumption that they have been missed by the compiler of a particular directory – quite possible in the early years of the sample. Also shown are first and last entries of distillers. Another complication, apart from the different spellings in this source, is the frequent change of partnerships recorded in the entries and sometimes confirmed in press notices. For example, an entry in the *Felix Farley's Bristol Journal* of 30 March 1793 (p.3b) announces the dissolution of a partnership of Thomas Cross, William Cross, Thomas Cole, William Parry the younger and Martin Petrie – none of whom are listed in the 1793 directory! However, the vibrancy of the industry in the 1780s, as mentioned above, is reflected in the 31 distillers active in 1785. Thereafter, there is a steady decline in numbers with fewer new entrants to the trade after 1795. In the years prior to 1815, some stability has been reached with just four distilleries, only one of which produced primary spirit.

More detailed information on the distillers and rectifiers of Bristol – but not Bath – is available in the detailed accounts presented to the House of Commons from HM Customs and Excise in the nineteenth century. Reliance on the duty raised from distillers was such – Michael Castle of the Cheese Lane Distillery paid the then massive sum of £55,611 6s 6d in the year ended 5 January 1803 – that a large local excise establishment took great pains to maximise revenue and seek out illicit distilleries. Newspaper reports, however, suggest that evasion was not uncommon: in December 1778, officers discovered a large quantity of spirits 'in a private unentered place' at a distillery 'near this city' and had to call the militia to their assistance (*FFBJ* 5 December 1778, p.3a). Sales of seized spirits, feints, molasses and equipment were also common (*FFBJ* 22 April 1786, p.3c). Table 2 shows the number of gallons of corn wash distilled in the Bristol and Bedminster distilleries between 1790 and 1805, by which date only one malt distillery remained in production. Only three malt distilleries were still active by 1790, Naylor becoming Castle & Co by 1793 and Parry, Cole & Co taking over the Bedminster Distillery from Cross & Co. Nehemiah Bartley's 'Capital Malt Distillery' and piggery for 1,000 hogs on Temple Backs had been offered for sale in 1792 (*FFBJ* 8 September 1792, p.3c).

The two surviving plants are also listed in a separate return of malt distilleries of 1802-3; both had three stills and produced over 200,000 gallons of plain spirit, Castle's output being slightly greater. This series of returns also includes a list of all English rectifying distilleries, and those in Bristol are summarised in Table 3. The addresses have been added to the latter as they appear in local directories of the time. Variations in the numbers of stills and their capacity is clearly indicated here, as are the significant differences in the scale of production. The proportion of raw spirits received into the Bristol rectifiers from the primary distillers and sent out is also shown – just under 10% of output in England.

Table 1 — Numbers of Distillers in Bristol from National and Provincial Directories 1775-1815

Year	Number of distillers listed in Directory	Number listed in previous/subsequent Directories	Estimated total of active firms	First entry	Last entry
1775	29	-	29	-	14
1783	20	6	26	11	5
1785	31	-	31	11	15
1787	16	4	20	4	3
1792	18	3	21	4	2
1793	15	4	19	-	3
1793/4	19	3	22	6	6
1795	15	4	19	3	3
1797	13	4	17	1	1
1798	10	6	16	0	1
1799/1800	12	3	15	0	0
1801	12	4	16	1	3
1803	14	1	15	1	1
1805	14	-	14	2	3
1806	9	2	11	0	1
1807	9	1	10	0	1
1808	9	-	9	0	1
1809	8	-	8	0	1
1810	7	-	7	0	0
1811	7	-	7	0	1
1812	6	-	6	0	1
1813	4	1	6	0	1
1814	4	-	4	0	0
1815	3	1	4	0	0

Sources
Sketchley's *Bristol Directory* (1775)
Bailey's *Western and Midland Directory* (1783)
Bailey's *Bristol and Bath Directory* (1787)
Browne's *Bristol Directory* (1785)
Reed's *New Bristol Directory* (1792)
Universal British Directory, Vol.11 (1793)
Matthews's *New Bristol Directory* (1793-4, 1795, 1797)
Matthews's *Complete Bristol Directory* (1798, 1799-1800, 1801)
Mathews's *Complete Bristol Directory* (1803, 1805, 1806, 1807, 1808, 1809, 1810, 1811, 1812)
Mathews's *Annual Bristol Directory* (1813, 1814, 1815)

Table 2 Number of Gallons of Corn Wash distilled in the years ending 5 July from 1790-1805 in the Bristol and Bedminster Distilleries

Distillery	1790	1791	1792	1793	1794	1795	1796	1797
John Naylor, Bristol	600767	797489	902126					
Nehemiah Bartley, Bristol	12430							
Castle & Co, Bristol				1087107	1261953	1497747	-	763236
Cross & Co, Bedminster	1431312	1511084	1316770	1231695				
Parry & Co, Bedminster						1365908	1425840	-	764575
Total wash distilled in England in selected years with the percentage of this produced in the Bristol & Bedminster distilleries	13136252 15.6%					28222473 10.36%		

Distillery	1798	1799	1800	1801	1802	1803	1804	1805
John Naylor, Bristol								
Nehemiah Bartley, Bristol								
Castle & Co, Bristol	1078503	1170305	1678851	-	916426	1481492	1312788	1702630
Cross & Co, Bedminster								
Parry & Co, Bedminster	834322	999167	1519246	5723	898536	892531	532382	
Total wash distilled in England in selected years with the percentage of this produced in the Bristol & Bedminster distilleries			26088033 12.3%					17166439 9.9%

Notes: Distilling from corn prohibited 10 July 1795-1 February 1797 and 8 December 1800-1 January 1802

Source: *British Parliamentary Papers* 1821 (335)

Table 3   Rectifying Distilleries in Bristol 1802-3, Number of Stills, Contents, Raw Spirits Received, Rectified Spirits Sent Out and Stock in Hand

Distillery	Number of Stills	Total content in Gallons	Raw Spirits Received	Rectified Spirits Sent Out	Raw Spirit in Stock in 1802	Rectified Spirits in Stock in January 1802
Robert Castle, Milk Street	5	5383	196361	292548	-	14410
Thomas Wigan, 138 Redcliff Street	2	737	35891	47095	50	1403
William Wanklin, King Street	2	609	339	109	65	123
Henry Crawley, 53 Broad Street	2	330	-	1314	-	128
John Banister, Broadmead	1	180	128	631	-	164
Timothy Cassin, Redcliff Hill	2	719	1714	1283	-	-
Joseph Dyer & Co., 98 Redcliff Street	2	633	12898	18988	39	1307
Joshua Powell, 28 Redcliff Street	3	1296	19537	29411	-	1474
William Moxham, Thomas Street	2	414	3530	3146	-	1292
Thomas Cale, Redcliff Hill	4	4122	103026	145556	1718	9711
Mary Fry, 104 Redcliff Street	left off		2950	1106	-	2592
Samuel Edwards, 24 Redcliff Street	2	1224	39710	55624	-	2651
Total in England			4958009 (8.4%)	6982504 (8.5%)	18531	346616

Sources: 1) *British Parliamentary Papers* 1802-3 (122)
2) Bristol directories as noted in Table 1

The final table (4 overleaf) has been compiled from a selection of surviving fire insurance policies from the period 1780-1786. These show the partners involved in the distilleries, their location, the value of the buildings and the combined value of stock and utensils. Policies often included a house for the owner, or one of the partners, and this has been excluded from the table, together with other property which does not appear to be linked with the working of the distillery. A division of fixed (i.e. plant and machinery) and circulating capital is not possible from these valuations but it is clear that the trade required very large sums of the latter and this is, possibly, the reason why Bristol distillers faced bankruptcy when bad harvests interrupted production.

Table 4 **A Selection of Fire Insurance Valuations of Bristol Distilleries 1780-1786**

Date of Policy with volume & policy number	Distiller	Address	Value of Buildings	Value of Stock & Utensils	Total Sum Insured
19 October 1780 287/434291	William Shorland John Wright & Wm. Sanders	Cheese Lane, St Philips	1,000	2,000	3,000
7 October 1780 287/433797	Christopher Load Thomas Gee & John Gee	Redcliffe Street	-	1,000	1,000
29 September 1780 287/433386	Joshua James	Stokes Croft	900	2,600	3,500
12 December 1780 288/436549	Gawin Allanson James Cross & John Naylor	Hawkins Lane	-	3,000	3,000
19 September 1780 287/433108	Joseph George Pedley	King Street	1,500	2,500	4,000
18 August 1783 314/481748	Joshua James	Stokes Croft	2,000	8,000	10,000
26 December 1783 319/487371	Joseph Dyer Butler Symons	Broad Mead		1,400	1,400
26 December 1783 319/487370	Nehemiah Bartley Samuel Worrall	Temple Back	700	8,600	9,300
13 January 1784 319/488369	Wm. Shorland John Wright & John Frear	Queen Street	-	1,000	1,000
14 April 1784 321/491676	Nehemiah Bartley John Jackson George Ewbank	Lewin's Mead	400	-	400
14 April 1784 321/491677	Nehemiah Bartley Wm. Moore & Philip George	Lewin's Mead	-	1,400	1,400
14 April 1784 321/491678	Robert Castle Rowland Williams John Ames & John Naylor	Milk Street	800	4,800	5,600
1 July 1784 322/494105	Joseph Hughes	Redcliffe Street	550	1,800	2,350
4 January 1785 327/500053	Thomas Cave	Redcliffe Street	4,000	6,000	10,000
15 April 1785 328/503578	Thomas Harris Handlon McCracken Tho. Hooper & Tho. Williams	Stokes Croft		4,000	4,000
13 March 1786 335/516128	James Cross Senior	Bedminster	500	6,000	6,500
25 March 1786 12/97036	Rob. Kyle Hutcheson	Redcliffe Street		2,500	2,500

Sources: Department of Manuscripts, Guildhall Library, Aldermanbury, London: registers of Sun Fire Office (MS 11936) apart from Rob. Kyle Hutcheson (12/97036) which is from the Royal Exchange Assurance registers (MS 7253)

Some Lost Distilleries in Bristol and Bath

In the absence of business records, maps and illustrations, the most graphic picture of the size and complexity of these 'lost' distilleries – and something of their history – can be pieced together from newspaper files, although these often dealt with businesses that had failed or were being sold at the time. Four examples have been chosen from Bristol and one from Bath.

The Jacobs Wells Distillery: Langdon & Hetling

The Jacobs Wells Distillery in Clifton and its owners, Messrs Langdon & Hetling, do not appear in the directories. A report of an accident in 1776 'at the new distil house at Jacobs Well' appears in the local press so it is likely that this started work after publication of Sketchley's directory (*FFBJ* 22 June 1776, p.3d).

The distillery experienced difficulties with HM Commissioners of Excise who ordered a sale of equipment and 'spirits lately seized at the Distillery at Jacobs Well' in 1779 (*FFBJ* 20 March 1779, p.3c). By early September, the creditors of another partner, Samuel Atlee, 'late of Jacobs Well Distillery' were summoned to discuss his affairs and the distillery was also offered for sale at auction at this time (*FFBJ* 4 September 1779, p.1d). These sale details provide a valuable guide as to the equipment and capacity of the distillery which was equipped to use malt or molasses: this included 3 large stills with worm tubs and worms 'of extraordinary size', a mashing tun of 35 quarters capacity, a large furnace for brewing, a chain wort pump worked by a fire engine (i.e. a steam engine), a large mill with an overshot wheel which worked two pairs of stones, capable of grinding 'all or most of the corn made use of in the house', a bolting mill for flour and two fire engines – one much larger and only 'made use of for trial'. On the steep slope above the yard was a large reservoir supplied by water from the small engine. The usual coolers and backs are noted together with stores, a dwelling and counting house (*FFBJ* 4 September 1779, p.2c).

In 1780, a field in Westbury-on-Trym in the possession of the estate of E. Langdon & Co. was offered at auction – upwards of £800 had been spent on building a piggery, dwelling house and sties, capable of containing 1000 pigs 'and is very fit for a distiller …' (*FFBJ* 1 July 1780, p.26).

Sale details of the Jacobs Wells Distillery, from Felix Farley's Bristol Journal, *25 September 1779.*

The sale of the distillery continued to 1785 when 28 washbacks, 10 under and spirit backs, a copper still and head (1000 galls) and a copper and lead furnace of the same size, a fire engine with a copper boiler and wort pump, a corn and dressing mill, water wheel and tin, lead and brass pipes and cocks were offered for sale (*FFBJ* 5 April 1783, p.2c). A final dividend on the estate of the company was to be paid in the summer of 1796 (*FFBJ* 30 July 1796, p.3d). A development application has recently been made for this site, now 19 Jacobs Wells Road in Bristol.

The Stokes Croft Distillery: Joshua James

Joshua James of the Stokes Croft Distillery is listed in directories between 1775 and 1785. Newspaper reports refer to partnerships with a Mr Perkins in 1776 (*FFBJ* 14 September 1776, p.3d), and Robert Jewkes (*FFBJ* 29 January 1780, p.2d).

The distillery was rebuilt after a 'dreadful fire' in March 1783 which destroyed the premises and several thousands of gallons of spirits but a large quantity destined for export and the stills and worms escaped the blaze (*FFBJ* 8 March 1783, p.36). It seems that it had been enlarged – the insurance policy of 1781 valued the distillery, brewhouse and adjoining buildings at £900 and utensils and stock at £2,600. By 1784, after the fire, he had insured his engine house, brewhouse, stillhouse, rectifying house and warehouses for £2,000 and utensils and stock at £8,000.

James's bankruptcy was announced in 1785 (*FFBJ* 9 April 1785, p.3b, *GM* 1785, p.329) and the distillery put up at auction early in 1786. Matters took some time to resolve and his creditors were still meeting some 19 years later (*FFBJ* 13 October 1804, p.3b).

Sales details of February 1786 describe the concern as 'capital malt and molasses distilleries lately erected and built … in a most substantial manner, and in work only two years'. Included was a 'compleat patent fire engine … almost new, with an iron cylinder 30 inches in diameter, capable of raising 9,000 gallons of water in one hour, which may be applied to the general work of the house'. Other plant included 'two brewing coppers with machines', (of 5,000 gallons and 5,500 gallons capacity) with connections to an oval mash tun that could hold 14,000 gallons with an underback of 3,300 gallons capacity. The pump was worked by the fire engine. Three coolers – two of 110ft x 30ft x 6in deep – were all connected by lead pipes and brass cocks. There were 14 oval backs in the same room, each taking upwards of 8,600 gallons and materials for making several more in a room available for expansion. The wash still contained 8,000 gallons, exclusive of the head, and was connected to a worm of part pewter and part copper in a tub of 34,350 gallons. The low wines still contained 2,000 gallons with a pewter worm and tub containing 12,000 gallons. The three rectifying stills held about 1,100; 800 and 60 gallons and the spirit still 1,800 gallons. There were eight underbacks for the low wines, spirits, feints and brandy and a spent wash back (5,500 gallons) all with lead pipes, brass cocks and pumps with brass chambers. Three barm backs were of 1,500 gallons capacity.

A large cellar was used as an export warehouse and there were two other large warehouses. James also had an interest in land on a long lease at Horfield, about a mile

A drawing by Samuel Loxton from Bristol: As it was – and as it is, *by Frederick Stone. This shows the Floating Harbour in Bristol, near St Philip's Bridge, c.1904. On the right, opposite the cranes of the railway sidings, can be seen the waterside buildings of Castle and Naylor's Bristol Distillery and its storage tanks.*

away, with sties and buildings for keeping 650 hogs, a dwelling house for servants and 40 acres of land. There was also a dwelling house, stable, garden and malthouse with a malt mill and conveniences for making malt in Stokes Croft, a small distance from the distillery. A less-detailed advert appeared in November indicating little or limited sale of the plant (*FFBJ* 4 February 1786, p.2c, and 4 November 1786, p.1c).

The buildings were again offered for sale in 1789 (*FFBJ* 19 September 1789, p.3b) and in May of the following year plans were announced to convert the dwelling house to a Magdalen hospital (*FFBJ* 29 May 1790, p.3e). Later press notices concern an estate called Southmead in Westbury-on-Trym which James had contracted to purchase (*FFBJ* 6 May 1797, p.3b). The sale was still being advertised in 1798 after his death (*FFBJ* 29 September 1798, p.1d) and again in 1803 (*FFBJ* 1 October 1803, p.2c).

The Hawkins Lane Distillery, Temple Cross: Cross, Allanson & Co.

The distillery of James Cross, Gawin Allanson and John Naylor is included in the insurance details in Table 4. This partnership was dissolved in June 1786, James Cross leaving to continue his own business at the Bedminster Distillery (*FFBJ* 24 June 1786, p.2e). The distillery and the materials of the pig sties 'now standing without Temple Gate' were advertised for sale a little earlier (*FFBJ* 10 June 1786, p.1c & p.1d).

The distillery was located on the eastward side of Hawkins Lane and extended to the River Avon. Plant details included are somewhat briefer than for some of the other distilleries but include a 'new-erected fire engine' together with the customary equipment for a malt distillery. Other properties included a dwelling house fronting Temple Street and a malthouse nearby in the Great Garden in Temple parish. Also included in this lot were 'certain erections and buildings ... new used as hog-styes', capable of receiving and fattening 700 hogs. It is not clear whether these are those mentioned above or additional capacity. The rectifying house in the second lot was situated on the opposite side of Hawkins Lane and also extended to the river. The fire engine in the malt distillery supplied this building with water. Other properties in this lot included warehouses and granaries and dwelling houses on Hawkins Lane and Tucker Street.

Later sales notices describe the premises as a brewery 'but now and for several years past converted into a distillery by Cross, Allanson & Co' (*FFBJ* 18 November 1786, p.3c). The site is now included in the former Bristol Brewery premises which is currently under development.

The Bedminster Distillery: Thomas Easton & Co., James Cross & Co., Parry, Cole & Co.

Parts of this distillery were in St Mary Redcliffe parish and part in Bedminster and it was active between the late 1770s and 1805. The site began life as the short-lived Bristol Porter Brewhouse of Attwood & White, which was offered for sale in 1778 (*FFBJ* 21 March 1778, p.2c). It was subsequently rebuilt and extended in the ownership of at least three partnerships.

A number of sales details appear in the press over the years, the first in the autumn of 1779 when the distillery was described as 'a new, large, commodious, and well constituted distill house'. The three-acre site housed five large and small stills with worms and tubs, one of which was of 2,000 gallons capacity. The mash tun could process 30 quarters of grain and the coolers were from 63 to 83 ft long and 13 to 18 ft wide. Ten backs were situated in a separate room with several not yet finished. There was a large new fire engine 'erected as a plan to answer different purposes for the use of the house' and a large well 'sunk at very great expence'. Facilities for use of the waste included sties for 600 hogs (*FFBJ* 23 October 1779, p.2d). The distillery first operated between issues of directories but seems to be that of William Hull and Matthew Wayne. A later advertisement for the sale of building materials suggests that it was built 'three years ago' (*FFBJ* 21 April 1781, p.1d)

The distillery was later taken on by the partnership of Thomas Easton, Thomas Cave and Samuel Edwards and again offered for sale when this was dissolved (*FFBJ* 15 January 1785, p.1c). It is described as 80ft broad by 130ft deep and had been used for only two seasons. The plant included a fire engine, 'almost new', with a copper furnace of 3,000 gallons and an iron cylinder of 24in diameter with two pumps capable of raising 22,000 gallons of water in one hour. There were two brewing coppers of 6,000 and 2,400 gallons capacity. The mash tun was of 23ft diameter and 6ft 6in deep with a pump worked by the fire engine to an underback. There were six coolers, the largest being 101ft 6in long by 16ft 3in and 6in deep. The capacity of the back rooms had also been increased with 13 square wash backs in one room and 6 round backs, iron-hooped, containing about 4,500 gallons each and another four which could take 10,000 gallons. All had brass cocks and lead pipes for charging and discharging.

The wash still was of 6,500 gallons and the low wines still 3,300 gallons, both with extensive worm and tubs. There were four oak underbacks for low wines, spirits, feints and brandy with pumps to convey the spirits to the store pieces. These were housed in a large warehouse with three pieces of 3,000 gallons and another of 2,000. There was also a two-horse mill to work the malt mill, which could also raise water if the steam engine was out of order. Other plant included six large corn lofts (to hold corn for two months), five backs for spirit wash, conveyed to the hog sties by the fire engine through an underground pipe to another back of 4,000 gallons that commands every stye. The rectifying house was in the parish of Bedminster and had two stills (2,000 and 1,000 gallons) with worms and four underbacks with pumps (*FFBJ* 7 May 1785, p.1d).

The distillery then passed to James Cross senior, whose insurance valuation is included in Table 4 with a total value of £6,500 for the distillery buildings, stock and utensils. The distillery was at work in 1786, when the parishioners of Redcliffe met to complain at the nuisance of the pig sties and the 'shot work', presumably the shot tower on Redcliffe Hill (*FFBJ* 25 November 1786, p.36). The excise accounts summarised in Table 2 show Parry & Co., also known as Parry, Cole & Co., had taken over from Cross by July 1794.

The final disposal of the distillery was a protracted affair, beginning when Cole offered it for private sale in mid January 1805. The distillery had been badly damaged by fire in May 1799 after a still head had blown off and only the walls, and some spirits stored in a cellar, survived (*FFBJ* 18 May 1799, p.3c). The 1805 details describe the premises as newly-erected. The plant included a steam engine capable of grinding all the corn and pumping all of the liquor used in the house from a well and a mill with four pairs of stones. The building now measured 125ft at the front and 168ft in depth. Detached from this was a cooper's yard and shed, a large warehouse with store pieces to hold upwards of 50,000 gallons and another to hold coal and empty puncheons. There was stabling for 20 horses and sties for 1,800 hogs and other sheds for casks and grains.

The rectifying house had four stills and the necessary worms and underbacks. Warehouses in Redcliffe Street extended back to the river Avon, which had been built by Thomas Cave for a malt distillery and used by Parry, Cole for corn lofts. Also available for a bacon factory near the Bedminster Turnpike with lofts for straw and wood, a dry cellar, stores and drying houses. The cellar was capable of storing 2,000 'hogs of bacon' and 300 hogs could be cured per week (*FFBJ* 12 January 1805, p.2c).

Later advertisements appeared in 1810 and 1821, the latter still referring to the premises as 'late the malt distillery of Messrs Parry, Cole & Co.' In 1821, there were three warehouses and a steam engine, corn mill, and lofts and warehouses that were then let to Michael Castle on an annual basis (*FFBJ* 10 March 1821, p.2c).

The Walcot Distillery, Bath: William Hetling & Co.

The absence of distillers in Bath from directories, excise accounts and insurance policies suggests that they did not exist here but the growing city provided a market for spirits and a number of dealers – and a few distillers – appear in local newspaper files. These include Thomas Maggs, taking over from James Biggs in Westgate Street and James Morse, a 'Chymist and Distiller', who produced 'rich cordial waters' made with rum and brandy (*BJ* 30 June 1746, p.2c and 25 May 1747, p.4a). Later in the century a meeting of the creditors of the distillers Hosier & Tunstal , 'late of Walcot', was called in 1785 (*BC* 13 October 1785, p.1c). The Northgate Brewery of Samuel Sayce also distilled: Sayce is listed as a brewer and brandy merchant in late-eighteenth-century directories and was reported as building on Bathwick meadow to fatten oxen with grains from his porter brewery and distillery in 1786 (*BC* 28 December 1786, p.3d).

Sayce's concern was clearly of some size but the only evidence of a concern to rival those in Bristol was the short-lived venture of William Hetling and Samuel Atlee noted above as partners in the ill-fated Jacobs Wells distillery in Clifton. Thomas Cave, the prominent Bristol distiller, had withdrawn from their partnership at the Walcot distillery in 1780 (*BC* 26 October 1780, p.3c). The distillery was still in operation shortly after this as complaints concerning the nuisance arising from keeping pigs at the new distillery were not upheld at Wells Assizes (*FFBJ* 10 August 1782, p.3d) but the bankruptcy of the concern was announced soon after (*BC* 19 September 1782, p.1c and *GM* November 1782, p.554). The distillery was put up for auction in July 1783. It consisted of a large brew-house with mashing vat to mash 70 quarters of malt, two coppers (about 2,000 gallons capacity each), a malt mill, still house with two stills (3,000 and 1,000 gallons), worm tubs, three large backs (40,000 gallons) and a rectifying house with two rectifying stills and worms. Also described is a back house with 14 backs (13,000 gallons) together with coolers, pumps and an engine house with a horse engine to pump water. The hog yard and sties could hold 1,000 pigs and was equipped with a spent-wash back to hold 40,000 gallons and an engine to serve the sties. This was an integrated plant with slaughter, salting and drying houses, a melting house, cask house and cooperage. A dwelling house with counting house and garden completed the plant (*BJ* 7 July 1783, p.4c).

The distillery was advertised on a number of subsequent occasions, the notice of August 1784 suggesting that it was 'well calculated for a brewery' and would be 'taken to pieces if not sold' (*BC* 26 August 1784, p.2c). In early 1785, some utensils were offered for sale, including the horse pump. A final dividend of William Hetling's estate and effects was announced in 1792 (*FFBJ* 5 May 1792, p.3d). Hetling was, apparently, rescued from his creditors by his wife's affluent family (anon 1986, p.8)

No evidence has come to light on later ventures in Bath – in 1797 an advertisement appeared for a partner willing to advance £2-3,000 to join a London distiller who intended 'entering into the same business near Bath' (*BJ* 10 April 1797, p.1c). Some 25 years later, an advertisement for the Albion Brewery in the city suggested that a distillery might be added 'at an easy expence' but this did not happen (*BM* 21 June 1823, p.1).

The End of Distilling in Bristol

After 1815, the Cheese Lane malt distillery was to continue in business and maintained the industry's record of innovation by installing the first Boulton & Watt steam engine in the city in 1793 (Bone, 1995). Under T.H. Board, it became an early limited company in 1862 and survived as an independent firm until 1917, when it became part of the Distillers Company Limited (Weir 1995, p.309). The distillery features briefly in Alfred Barnard's classic guide to the U.K.'s whisky distilleries, when it produced plain spirit for the rectifiers and grain whisky for sale to the blenders of Scotch and Irish whiskies (Barnard 1887, pp.451-5). The business ended in the 1980s as a yeast factory and the old Cheese Lane (now Avon Street) site lies in a development area. The last of the Bristol rectifiers, Castles of Milk Street, was to close in 1869 when the business relocated to Liverpool (*BTM* 1 May 1869, p.1c).

There is now no discernible trace of an industry that was in the forefront of technical innovation and large-scale production. However, careful work on rate books, deeds, and rentals should be able to identify most of the old distillery sites and add these to the Historic Environment Records that are consulted when development applications are made. A recent article in the *Journal of the Association for Industrial Archaeology*, of which L.T.C. Rolt was the first president, demonstrates the potential for archaeological excavation of a long-lost distillery prior to development of a site in Edinburgh (Heawood, 2009).

Some of the social problems associated with drinking cheap mass-produced spirit still, of course, remain in Bristol, Bath and elsewhere in the U.K. and recent books by Dillon and Warner draw parallels between the age of gin and the widespread use of drugs in recent times. Vodka, a much-less sophisticated product than gin, has now taken pride of place as the binge drinker's spirit of choice. 'Madam Geneva' still has her followers but has now made the long journey to respectability, via the Georgian slum and the Victorian gin palace, to the cocktail bars and drinks trays of polite society. The recent revival of small-scale production of cider brandy apart, Plymouth Dry Gin is now the only traditional spirit distilled, or rectified, in the West of England.

Abbreviations of Newspapers and Serials Consulted

BC   *Bath Chronicle*
BJ   *Bath Journal*
BM   *Bristol Mirror*
BTM   *Bristol Times and Mirror*
FFBJ   *Felix Farley's Bristol Journal*
GM   *Gentleman's Magazine*

Bibliography

anon., 1986, 'The quest for Hetling', *Mayor's Guides Newsletter*, Bath, 28 November 1986.
Ashton, T. S., 1972, *An Economic History of England: the 18th century*, London: Methuen.
Barnard, A., 1887, *The Whisky Distilleries of the United Kingdom*, Harpers Weekly Gazette, London, reprinted in facsimile, with additions, by Birlinn, 2008.

Barrett, W., 1789, *The History and Antiquities of the City of Bristol*, Bristol: William Pine, reprinted in facsimile by Humanities Press, 1982.

Bateman, J., 1840, *The Excise Officer's Manual and Improved Practical Gauger*, London: A. Maxwell.

Berlin, M., 1996, *The Worshipful Company of Distillers: a short history*, Chichester: Phillimore.

Bone, M., 1995, 'Boulton & Watt steam engines in Bristol and Bath', *BIAS Journal*, 28, pp.24-9.

Coleman, D.C., 1977, *The Economy of England 1450-1750*, Oxford: Oxford University Press.

Dillon, P., 2002, *The Much-Lamented Death of Madam Geneva: the eighteenth-century gin craze*, London: Review.

Dodd, G., 1843, *Days at the Factories*, London: Charles Knight, reprinted in facsimile by EP Publishing 1967.

George, E. & George, S., 2005, *Bristol Probate Inventories Part II: 1657-1689*, Bristol: Bristol Record Society.

George, E. & George, S., 2008, *Bristol Probate Inventories Part III: 1690-1804*, Bristol: Bristol Record Society.

Harper, W.T., 1999, *Origins and Rise of the British Distillery*, Lewiston: Edwin Mellen Press.

Heath G., 1799, *The New Bristol Guide*, Bristol: R.Edwards.

Heawood, R., 2009, 'Excavations at Lochrin Distillery, Edinburgh', *Industrial Archaeology Review*, XXXI(i), pp.34-53.

Latimer, J., 1893, *The Annals of Bristol in the Eighteenth Century*, Bristol: the author.

Mathias, P., 1979, *The Transformation of England*, London: Methuen.

Rolt, L.T.C., 1970, *Victorian Engineering*, London: Allen Lane The Penguin Press.

Thirsk, J., 1978, *Economic Policy and Projects: the development of a consumer society in early modern England*, Oxford: Oxford University Press.

Walton, K., 1986, *An Inventory of 1710 from Dyrham Park*, London: Furniture History Society.

Warner, J., 2003, *Craze: gin and debauchery in an age of reason*, London: Profile Books.

Weir, R.B., 1995, *The History of the Distillers Company 1877-1939*, Oxford: Clarendon Press.

4: Managing a West Indian Sugar Estate: John Pinney and the Island of Nevis

Owen Ward

Sugar became an important commodity of trade and industry in Bristol in the seventeenth and eighteenth centuries. At the time of the events recounted here plantations in the West Indies were offering satisfying profits, but only for determined proprietors. Some of these had been prosperous enough in other fields to be able to invest a sizeable working capital in a concern worked up by others: a few, on the other hand, had to sustain a testing period of hard-won development before the returns amounted to a comfortable sum.

The plantation business, and the fortune, of the eighteenth century Pinney family of Bristol was in large part the result of an unhappy accident. In 1685 the hothead of the family, Azariah, was so careless as to get involved on the wrong side in the foredoomed attempt by the Duke of Monmouth to grasp his throne from James II. Even more carelessly he was caught, tried and imprisoned to await execution. But he had a level-headed, prosperous brother Nathaniel who was an independent trader with shares in the Royal African Company and was thereby entitled to indenture servants to go out, in slightly less discomfort than the slaves (which he was also entitled to supply), to serve him in the West Indies. This is how he came to spend £100 purchasing Azariah to go out as his servant, although he could not have expected or wanted such a go-ahead individualist to spend all of his committed years in a subservient role.

Azariah settled on the small island of Nevis, which was at that time the seat of British West Indies colonial government. Throughout the eighteenth century it was an island of sugar plantations owned by British landlords, though they were often absent, and the plantations were worked by slaves and indentured servants with some salaried technicians and overseers. Once he had reached Nevis with a few pounds in his pocket, Azariah worked at making a success as a salesman for the family business of producing and selling lace, receiving payment in plantation sugar so that by sending cargoes of sugar back to his brother he was able to pay off his bond and launch out on his own account as sugar plantation manager and general merchant.

The remittances of sugar from his estates soon came to be worth several thousand pounds at a time, and he returned to England in 1719. He also profited largely as a West India merchant, supplying all kinds of provisions and plantation equipment for the island's residents in exchange for remittances of sugar from their estates. As usually happened, his island plantations were less successful after he left them in the hands of managers, but his son, John Pinney, succeeded him and went out to Nevis to stimulate a revival in the plantations' fortunes. His son in turn took over the business, but left the island estates in the hands of agents, so that a decline was inevitable, until John Pretor, a cousin, was asked to join the family business and take the family name in 1762.

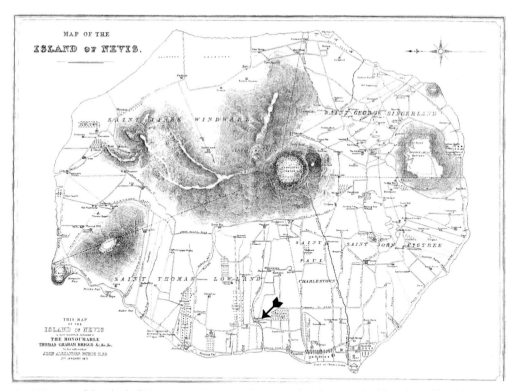

An 1871 map of the island of Nevis. Pinney's estate can be seen a short distance inland from the middle of the Western coast, indicated here by an arrow at the bottom of the print.

John Pretor Pinney diligently set about reviving the business so effectively that he built up a family fortune by the time of his retirement in 1789. He too found it essential to spend periods on Nevis to ensure the proper management of his plantations, but he also conducted an efficient and profitable trade as a sugar factor from the office in Bristol which became his home in Great George Street. One, rather exceptional, customer who became resident in the West Indies was Ulysses Lynch. The story of Ulysses Lynch is unfortunately not a completely satisfying one. It has no real beginning and no conclusive end. That is mainly because the records now housed amongst the Special Collections of the University of Bristol Library are concerned with the work which the Pinneys did, and only incidentally with such 'friends', as Pinney called his customers, as Ulysses Lynch.

The documents which have been especially studied for this particular episode in the affairs of the House include 30 of the large manuscript volumes of copy letters; but there are over 200 of them, besides hundreds of loose papers. The firm's letter books have been carefully investigated in the past by Professor Richard Pares of Edinburgh University, but at that time some of them were still scattered among the family's private holdings. His *West India Fortune* was published in 1950, though it is not now easily come by. Professor Pares has a couple of brief references to Ulysses Lynch in his book and these have been used to fill in gaps in the narrative.

Mr Lynch settles in the West Indies

On 16 November 1787 John Pretor Pinney, by that time head of the House, wrote a letter to William Coker, a member of a West of England family long established on the island of Nevis, and Pinney's agent there. In his letter Pinney informed Coker that he had entrusted Mr Lynch with 'the duplicate of the deeds' in order to take them with him to Nevis whither he was about to go and settle down. In a postscript the next day he added that he had also sent up-to-date newspapers 'under care of Ulysses Lynch esq.' down to the ship bound for the island .

The deeds will be those of Symond's estate in which Pinney was, albeit reluctantly, in process of investing with the hope of recovering the money which he had advanced for the provisions and materials which he had been supplying to it. The resident planter had been, as usual, expected to remit its sugar to Pinney for him to sell on commission so that he could recover the money on the goods which he had advanced, but it has to be said that even when Pinney entered it himself he found that this particular estate only got deeper into debt, and although he ran it efficiently he eventually recovered only part of what it owed him.

Initially Pinney obviously looked upon Lynch as a good and dependable friend. As he wrote at the end of 1794, Mr Lynch had 'lived with us on mutual terms of conviviality' while living in Bristol. It seems likely that the Pinneys' relationship with Mr Lynch's family was a long-standing one. The name of Lynch does suggest an Irish background, which is where the Pinneys could have made the Lynches' acquaintance. A century earlier John Pinney's predecessor, John Pinney the Preacher, visited Ireland twice, in 1663 and 1666, and then lived in Dublin between 1683 and 1688 after his non-conformity led to his eviction from his church in Dorset. One of his daughters was apparently already in Dublin, engaged in the family lace business. 'Preacher John' was extremely popular in Dublin, probably especially so amongst the English Protestants who had been settled in Ireland by their aristocratic English landowners, and the family connection with Ireland was certainly maintained for many years.

After his arrival on Nevis in 1787 Lynch set himself up in business under the name of Samuel Lynch & Co. though there is no other mention of a Samuel in the records so far scanned. Ulysses himself, however, moved across to the neighbouring, larger island of St Christopher, later known as St Kitts. By the end of January 1788 Pinney was addressing his letters to Lynch on St Christopher, and assuring him that 'were it in my power to do you real service, you would find no one of your friends more ready and happy in doing it'. But Lynch must have found the business of storekeeper in the West Indies rather tougher than he expected, no doubt discovering that the islands were already served by a tightly knit côterie of merchants. In August 1788 Pinney wrote saying that he was 'truly sorry to hear you have met with such oppositions and disappointments', but he went on to hope that the business would eventually prosper, and that meanwhile he would be 'particularly happy to serve you' in Bristol. At the same time Pinney had been advancing Lynch goods for resale as a friend but was forced to warn him that his business at Nevis stood 'indebted to me £35 9s. 7¾d. instead of there being a small balance in your favour' as Lynch had clearly hoped, or even expected. This small debt, precisely enumerated as was Pinney's inveterate way, would have been for goods supplied for Lynch for resale to planters on the island.

Such a service on Pinney's part was unusual if not unique because, as Pares emphasises, sugar factors like the Pinneys normally preferred not to deal with island storekeepers like Mr Lynch at all. For one thing they were in a way competitors in their own business of supplying goods to planters. Moreover, storekeepers often had little or no security to offer, since they normally had no land and no produce of their own. Repayment for the goods sent out to them had to be made in bills, or IOUs, on British firms which they had received from the islanders in payment for the goods supplied. Pinney then added a mere ½% to the storekeeper's account for his trouble in arranging to cash the bill. This small amount was nothing compared to the 2½% which Pinney obtained when he sold a planter's sugar for him, which was also a more secure source of income.

This is no doubt one reason why Pinney had not so far entered into any formal business relationship with Lynch, so that in February 1789 Lynch was hinting at finding a 'permanent correspondent' elsewhere. Pinney was not averse to Lynch's suggestion, but was nevertheless prepared to write and say that Lynch would find him 'willing and ready to enter into any connection with a Gentleman of your established character on terms that may be likely to prove convenient and advantageous, mutually to both parties'. Lynch was shortly expected to pay a return visit to England when Pinney would 'be happy to take you by the hand again'.

In spite of all this bonhomie there was a sour note struck in late August 1788. Pinney was replying to Dr Thomas Pym Weekes, his wife's half-brother, and a qualified doctor. At the time he was the resident manager on the estate which Pinney had inherited in the south of the island. Pinney's letter, as copied into his letter book, ran thus:

> [Your letter] hints that my friend Mr L. hath not conducted himself towards me in the manner you expected. I hope and cannot help believing, and so does your sister, that his conduct has been misrepresented – he certainly is incapable of being guilty of such duplicity. I am much obliged to you for the information, well knowing that it springs from the purest motive …

We are bound to deduce from later correspondence that 'Mr L' was none other than Mr Lynch. Unfortunately Weekes's letter has not been traced amongst the existing documents. Indeed the letters from correspondents in the West Indies, where they exist, remain to be analysed, so what they contain can only at present be surmised from the family's own replies to them as they appear in the firm's letter books.

Mr Lynch – the awkward customer

Pinney continued to try to do business with Lynch though the latter's orders were sometimes enigmatic. Among the items which Lynch ordered in September 1788 for example was 'shot': Pinney observed that he had 'mentioned no particular size therefore sent such as we imagine will suit'. He also asked for 'keys to take off wheels': but 'as you sent no sizes an ironmonger says that it would not be right to send them by guess as they might prove useless'. Another item was 'floor cloth': again 'you sent no dimensions nor could Mr[s] Lynch furnish us with any. We have therefore sent a middling sized one which we hope will answer' [how big is a middling one, one wonders?]. Mrs Lynch evidently remained in England: there was talk of her going out to join Ulysses on St Christopher, but we have no evidence that she ever did so.

In a letter written to Ulysses Lynch on 20 November 1789 John Pinney is found patiently explaining why it was that a particular order could not always be despatched as promptly as both sender and recipient might like. He must have been writing in reply to a complaint from Lynch, who was perhaps still finding things 'difficult' on the islands, because of the slow delivery of ordered goods. Moreover he seems to have been comparing his own lines of communication with those of the well-established merchants whose correspondents were in London, and who received their goods more quickly than he did his.

Pinney explained that Bristol merchants had to provide their own ships as there were so few available for hire. London factors could always charter a ship, though they may have had a share in one. Bristol merchants on the other hand always had to purchase their own ships which they then had to try to fill on both outward and homeward trips to cover the cost of the voyage. This might mean holding back a sailing to await more business, so delaying goods already destined for the journey.

In spite of his problems, Lynch's business did develop in some directions, so that by May 1790 we find him dealing with heavyweight items including, among other ironware, three or four sets of mill cases. These were substantial iron rollers, three of which were fitted vertically side by side in a sugar mill and rotated as a unit to crush the cane as it was fed through on one side of the central roller and back again on the other side of it. Because most of the mills at this date had originally been built of timber, even the rollers, they were not made to a standard size: the dimensions depended on what tree wood was available. When new fluted iron rollers replaced the old wooden ones they had to fit the elaborate, heavy framework already there, or else a new mill had to be built. Unfortunately, when sending for a set of mill cases in early 1790, and in spite of John Pinney's earlier admonitions, Lynch's attention to the detail in his orders had shown no signs of improving. On 5 May 1790 Pinney had to write back to him to say:

> we must be free enough to remark that there are some parts of your last order which appear rather obscure – particularly that relating to the mill cases. You order "1 sett of mill cases with everything complete agreeable to former order". Now on looking back to your former order we find it was for three setts of mill cases, of different sizes ...

Pinney promised to study the order again, and if he could make sense of it would 'send out the goods ... if we meet with a fit opportunity'.

As we have already noted, from time to time John Pinney was obliged to go to Nevis to superintend his estate there. When absent from the island he appointed agents to take charge of the business. They always began their tour of duty with exemplary enthusiasm, but their attention and conscientiousness gradually dwindled. Whereas their early reports

An early French engraving of a sugar mill driven by mules, from Sucrerie, *published by N. Bénard, Paris, 1784.*

were full of detail, and of criticism of the previous incumbent's laggardly attitude, the daily record in the plantation diary (of which there is one in the Bristol collection) became more and more fragmentary. The climate, and more especially the rum, took their toll on the managers' capabilities until their final legacy was no better than that of their predecessors.

It was because of this degeneracy of successive estate managers that from April to September 1790 John Pinney took himself to his island estate. Even while he was on Nevis he continued to write, sometimes at length, his business correspondence. Copies were kept in his letter books which seem to have crossed the Atlantic with him. In May 1790 he wrote in an attempt to placate Ulysses Lynch, a few miles away on the adjacent island of St Christopher, because Lynch had told him that 'two baskets of cheese were landed in bad order'. But as Pinney was so near at hand he was able to write again less than a fortnight later to say that the ship's captain had shown him the cheeses in question, and that, of the two 'only one was injured and that was broke in two, owing to its being put up too old which was no fault in the stowage'. Lynch had obviously been trying to lay the blame on Pinney's captain in his handling of the consignment, but Pinney said he 'had delivered the cheese in as good order as the nature of the goods would permit'. In any case, he went on, 'when I see you [which he presumably planned shortly to do] I will settle the amount agreeable to invoice in order to prevent a dispute upon so trifling a matter' and probably enjoyed the cheeses himself.

In the same letter Pinney referred to some tables which he had for sale, perhaps he had brought them with him on the boat as a speculation or in response to an aborted request from one of his 'friends'. He declared however that he would not 'think of selling them under £14 14s. 0d. sterling, the first cost in London'. If Lynch did not want to buy them at that price Pinney reckoned he would 'be able to procure a purchaser for them on [Nevis]'.

There was another quibble in September 1791 about 'spoilt' rolls of Penistone. This was a cheap, coarse woollen cloth, probably from Yorkshire where it originated in the town of that name. In this case it would have been supplied by the plantation owner to his slaves for them to make their own clothing. Lynch conveyed his concern, probably at the instigation of one of his customers, that 'the Penistones have a dark stripe added at each end of the roll' making a little bit of it useless. But it was pointed out to him that Penistones came in long rolls of 50 yards each, and that the stripe was necessarily worked in by the maker to guard against pilferage.

A disconcerting practice leads to dissension

The next spate of correspondence between John Pinney and Ulysses Lynch takes an unusual turn. In December 1791 Pinney wrote to Dr T.P. Weekes, whom he had by now left in charge of his estate on Nevis, to explain that he had 'received a letter from Mr Lynch, complaining that a parcel of windmill cases, sent out to him two or three years ago, do not answer the weight charged in invoice'. It appears from later correspondence that the cases concerned were a set of three sold to a Mr Stedman Rawlins, a planter on St Christopher. Mr Lynch's complaint bothered Pinney more than such letters often did. As he went on to write to Dr Weekes: '[my own mill cases] (now on board the *Hercules*) were cast by the same manufacturer [and] I request the favour of you to have them landed in [Charlestown] and carried to the King's scales, where you will see them weighed yourself'. He should then let Pinney have a chart showing the actual weight against the weight charged in the founder's invoice.

Pinney clearly felt unable to place any trust in the fractious Lynch, and besides his letter to Weekes he wrote two others. One was to Lynch, pointing out that he could not verify his allegation, because Lynch had omitted to specify at this point which size of mill cases were involved. At the same time Pinney took the unusual step of asking Lynch to return any mill cases which he had on hand via Pinney's own ship, the *Nevis* (there is a letter to the ship's captain advising him of this). The cases would be brought back to Bristol and afterwards returned to St Christopher, freight free, so that Pinney could 'have the satisfaction of seeing them weighed here in the presence of the maker'. He went on to say that the maker concerned enjoyed an impeccable reputation, but the overstatement of the weights on a number of items 'smacked not so much from inattention: but when so many appear deficient' it 'carries the appearance of settled, and intentional fraud, and the matter ought to be thoroughly investigated'.

Pinney also wrote, on the same day, to one Nicholas Richards, an attorney resident on St Christopher. He sent him a copy of the letter which Lynch had written for Richards to read for himself and asked him if 'there would be any indelicacy in applying to Mr Stedman Rawlins to get the exact dimensions in length and diameter of the three cases he purchased of Mr Lynch, as well as the exact weight of them'.

Lynch in his letter to Pinney had thought to specify the comparative weights of the consignment as a whole: the total weight of the cases on the invoice came to 29cwt 0qtrs 13lbs [3261lbs], but 'to my astonishment they weighed only 2791lbs, diff[eren]t 47lbs, and had the weights been new the difference would be more'. His expressed lack of confidence in the accuracy of the weights explains why Pinney had insisted that Dr Weekes should have Pinney's own cases, at present in transit, weighed by the King's weights as supplied to the Governor in the capital, Charlestown.

John Pinney's patience is tried

For a while Pinney continued to honour Lynch's orders for more ironware: a consignment of windmill parts, with all the individual weights carefully recorded, and amounting to a total bill of £66 0s. 7½d., was despatched in January 1792. But within a couple of months Pinney was refusing him further credit, and was 'highly pleased to find you have settled a correspondence in London or elsewhere to your entire satisfaction'. Which should have prefaced the closure of the book on Lynch – once he had settled his debts.

However by 2 April 1792 Pinney received from Lynch a letter which, replied Pinney, was 'more particular with the size and weights of the mill cases you complain of' so that he felt that he had to 'scrutinise into this extra ordinary and unexpected affair'. A couple of months later a further communication from Lynch must have expounded the problem so clearly that Pinney was prepared to confront the manufacturer and he was able at last to write back to Lynch that he had done so:

> ... we have come to a full explanation with the iron founder respecting the apparent deficiency in your different mill cases. Indeed ... a few words answered the purpose ... the cases are weighed <u>in the rough</u>, just as they come out of the moulds, and the workmanship charged at so much per cwt on that <u>rough weight</u> which ... appears in the invoices ... the <u>difference</u> ... is owing to what <u>comes off</u> and is lost in the turning, which must of course vary, even in cases of the same dimensions ... Mr Rawlins will readily see the propriety of returning to you what you have been induced to take off your account in consequence of his remonstrances, and you [will be able] to <u>sell</u> the rest on their proper terms. At the same time we must candidly confess we do not consider it as the most equitable way of charging, and we think it would be fairer to weigh the work <u>after</u> it is finished, and to add something more per cwt for workmanship, and under this impression <u>we</u> are determined to try to get that custom altered.

But, unhappily for Mr Lynch, he met his match in the forthright, but obstinate, Mr Stedman Rawlins. Pinney, as a man of honour, though a man of business, would have expected no less proper an attitude on the part of a fellow Englishman, and a fellow planter. But he was disappointed, and Rawlins refused to refund Lynch. It was not until March 1792 that Lynch admitted that his own perseverance had been ineffectual and he appealed to Pinney to support his claim for the return of the balance of the bill for the mill cases which was now seen to be correctly due to him.

Up to now it had not been stated in the correspondence who the iron founder was.

But in one of the account books is the following entry for 22 March 1793:

> Ulysses Lynch – paid for getting City Seal affixed to the deposition of John Winwood (Iron founder) relative to some Mill Cases sent Mr Lynch, 6s.6d., stamps for do. 2s.1d. … 8s.7d.

The charge for the City Seal, of course, was added to Lynch's debt, as he was advised in Pinney's covering letter of 30 March, a message which was delayed, along with the *Nevis*, which waited for a convoy for six weeks and more.

John Pinney, with Ulysses Lynch's contentious nature in mind, as well as the outstanding professional reputation enjoyed by John Winwood, was particularly reluctant to attribute any fault to the founder. Mr Lynch, however, was determined to pursue the matter of the mill cases, and correspondence from him on this subject continued to arrive in Bristol for another 18 months. Pinney replied patiently in September 1793 and again in March 1794. But in spite of continued negotiation with Winwood, suggesting for example that the founders should 'add a proportion of wastage as in London', John Pinney failed to upset the iron founder's 'established custom'. Eventually, in November 1794, Pinney had grown tired of Lynch's continued badgering. He wrote two long, careful letters, one to his attorney on St Christopher, Nicholas Richards, and one to Ulysses Lynch. They show how far, and how irrevocably, their relationship had deteriorated.

It appears that Lynch had urged John Pinney to take Winwood's to law: but Pinney refused to waste either his money or Lynch's on such 'a precious point of law'. He went further, and he had rarely been so outspoken in his correspondence:

> under all these circumstances … we must be candid enough to confess that we differ so much from you on our notion of business that we are anxious to have the connection finally closed between us and with your opinions we presume you must be equally desirous to bring it to a termination …

Mr Lynch's disturbing conduct; the Pinneys' ultimatum

On the same day Pinney wrote to Nicholas Richards:

> We have received this year several complaining letters as usual from Mr Lynch with a promise of a remittance of £200 instead of which he has only sent us £39 12s. 4d. which is not sufficient to pay the interest of his balance … we must beg the favour of you to endeavour to procure for us as large a remittance as you can. Our connection with Mr Lynch commenced more as friends than as merchants as he was settled in this city … so that perhaps more was expected on both sides than common and we trust that in our conduct he has experienced more than the usual indulgence.

Once the affair had been explained so clearly to the capable, and thoroughly reliable, Nicholas Richards, Pinney again expected to have little more trouble from Mr Lynch. But he was disappointed to find himself once more drawn into the affray in the summer of 1795. At this point he again had to write to Lynch demanding payment of his debts, now amounting to £1462 2s. 9d., and to Mr Richards asking him, perhaps injudiciously, to see to it 'by any manner you can'.

What manner this was did not become apparent until, over a year later, in October 1796, Pinney wrote to Richards:

> We are most sincerely concerned, as well as much surprised, at the treatment you have experienced from Mr Lynch: we feel therefore particularly happy, in having it in our power, to congratulate you on the fortunate result of the meeting with which you honoured that Gentleman; as it would have given us much real pain, had anything more disagreeable happened, in a rencontre which originated in your friendly zeal to serve us.

What occasioned this outburst of contrition? Here Richard Pares has found the answer. He writes that Lynch 'behaved so badly in Nevis that their attorney had to fight a duel with him'. Such an expedient was not so unusual when a gentleman, which Mr Lynch evidently now felt himself to be, decided that he had been accused of dishonesty, nor is it unusual to find no local record of a duel. It is reasonable to suppose that by 1796 the chosen weapons for duelling would be pistols. Flintlocks were being made specially for the purpose, in pairs, from 1770, and firearms were often included in the lists of goods sent to the West Indies, more especially in time of war. But flintlock pistols did not always fire, and it seems that because of this, or the questionable accuracy of their weapons and the combatants' inexperience, they both missed. And they may have deliberately avoided any outcome serious enough to have courted public disapproval. But if honour was satisfied as a consequence of the encounter, nothing else was, and especially not Lynch's debt to the House of Pinney.

It is no surprise to find that Pinney did not give up even when he 'retired' as the head of the House in 1789. As late as February 1798 there is a letter to James Stephens in London, who must have been Lynch's new correspondent, which begins 'Mr Pinney has communicated to us …'. So it must have been written by the new, young partners in the firm, Pinney's own son and two young Tobins. They carry John Pinney's torch well:

> We must be candid enough to say that after Mr Lynch's very unjustifiable behaviour to our attorney Mr Richards, which might have been attended with the most fatal consequences, we cannot think that Mr Lynch can feel himself the least entitled to any indulgence from us.

Over a year later, in April 1799, to the same gentleman in London, the partners wrote an answer to a letter 'the contents of which have not a little surprized us.' They refer to a proposition made by Lynch for a rescheduling of the payment of his debts by instalments. But they observed that he 'has made so many fair promises, and so often deceived us, that we have no reason to suppose he will be inclined to attend more rigidly to the arrangement he now proposes'.

They ended their correspondence with 'a determination to receive £100 … each on 1st Jan 1800, 1801, 1802 1803 and 1804' even though this meant giving up more than half the money owing to them. Richard Pares established from his painstaking study of Pinney's idiosyncratic accounting system that such indeed was all that the House ever recovered from this anomalous and irksome connection.

Envoi

The events described here exemplify a few of the problems experienced by absentee planters and West India merchants in the eighteenth century. The equipment which was ordered for the West Indian customers from the office in Bristol was rarely seen by the merchant himself before it was loaded on board ship. The offending mill cases in question had not been inspected by Pinney prior to lading, and it was one of the rare occasions when he was unable to persuade his supplier to improve his business methods. Much the same difficulty applied to the provisions, especially those supplied in bulk for the slaves. We read, for example, how disappointing were consignments of some miserable Severn herrings as a trial replacement for the usual fat ones which had to be ordered from Scotland. The beef was also sometimes complained of, including some sent by Ulysses' brother Marcus. Ulysses had no part in this particular transaction.

The absentee estate owner, as John Pinney was only too well aware, experienced the additional difficulty of maintaining a grip on the efficient running of a plantation. He was bound to delegate, to a manager who remained in the islands, the job of keeping the workforce as well as the machinery in good condition, and of ensuring that the sugars produced were in good merchantable state. Typically, his first reports received from the newly appointed manager served to reassure the owner that the work was in good hands, but the effects of readily available rum, often tainted with lead from the lining of the containers in which it was kept, would gradually obfuscate the agent's diligent intentions, until total neglect set in.

John Pinney's immediate successors struggled for a while with all these difficulties together with the added problem of dwindling profits from the sugar business until the island estates were eventually sold off to a neighbouring planter who was resident on the island and retained tight personal control, whilst the various members of the Pinney family settled down at home to enjoy their well-found English estates.

Select Bibliography
Cawton, Christopher J. and O'Shaughnessy, Andrew J.: 'Absentee control of sugar plantations in the British West Indies' in *Accounting and Business Research*, Vol.22, no.85, 1991, pp.33-45.
Hopton, Richard, *Pistols at dawn: a history of duelling*, London: Portrait, 2007.
Macinness, C.M., *Bristol: a gateway of Empire*, Newton Abbot: David and Charles, 1968, chapter XV.
Nuttall, Geoffrey F. (ed.), *Letters of John Pinney 1679-1699*, Oxford: Oxford University Press, 1939.
Pares, Richard, *A West India Fortune*, London: Longmans Green, London, 1950.
Torrens, Hugh R., 'Winwoods of Bristol: 1767-1788' in *Bristol Industrial Archaeological Society Journal*, no.13, 1980, pp.9-17.
Ward, Owen, 'Winwood & Co., Bristol' in *Bristol Industrial Archaeological Society Journal*, no.29, 1996, p.46.
Ward, Owen, 'The Pinneys and the Hon. Peter Jeffery's steam engines', in *Bristol Industrial Archaeological Society Journal*, no.39, 2007, pp.31-36.
Watts, David, *The West Indies: patterns of development, culture and environmental change since 1492*, Cambridge: Cambridge University Press, 1987/1990, pp.334-336, 547.

Acknowledgements
The author is grateful for the assistance of successive archivists working in the University of Bristol Special Collections. Archivists in Dublin and on Nevis and on St Kitts have been good enough to hunt, albeit without success, for any record of Ulysses Lynch before and after his association with the House of Pinney. There is still the possibility of a reference to him in the papers of James Stephens, merchant of London, if they can be found. The Centre for the History of Technology at the University of Bath has continued to encourage my endeavours.

5: James Nasmyth: Engineering Astronomer*

Angus Buchanan

James Nasmyth (1808-1890) is remembered primarily by historians of technology as a great mechanical engineer and as the inventor of the steam hammer. He enjoyed the unusual distinction of having his autobiography written for him by Samuel Smiles, and he was able to impress upon Smiles the importance of astronomy in the later part of his career, but for the most part historians who have given Nasmyth serious attention have chosen to ignore this aspect of his life. A.E. Musson, for instance, in the course of several excellent articles about Nasmyth, hardly mentions his astronomical aspirations, and John Cantrell, in his more recent study of the engineer, does not touch upon the subject at all.[1] Even L.T.C. Rolt, in his characteristically incisive review of the men who made the machines in the Industrial Revolution, *Tools for the Job*, does not comment on this aspect of Nasmyth's work: nor does he make anything of it in his study of *Victorian Engineering*.[2] Such neglect is a pity because, even though it cannot be claimed that Nasmyth attained the rank of being a leading astronomer, he nevertheless made some sterling contributions to astronomical observations of the sun and the moon, and won the respect of contemporary leaders of the profession. His career as an astronomer is thus at least an interesting footnote to the history of science and technology, and it is the aim of this essay to assess the nature of his achievement in this field.

Portrait of James Nasmyth, engraved by Charles Roberts. (frontispiece from James Nasmyth, Engineer: An Autobiography, *edited by Samuel Smiles; John Murray, London, 1885)*

Nasmyth was a Scotsman who, like so many of his countrymen during the period of rapid industrialisation in Britain, moved south in search of fame and fortune. Born in Edinburgh, into a well-established artistic family, he retired from a brilliant career as a mechanical engineer in 1856, at the age of 48, to pursue his gentlemanly hobbies of astronomy and photography, and he lived on to

---

* An earlier version of this paper was presented to the 18th International Congress of ICOHTEC, the International Committee for the History of Technology, held in Paris in July 1990, and subsequently published as 'Engineering Astronomers: The contribution of some practical men to the science of Astronomy', in Alexandre Herléa (Editor), *Science – Technology Relationships* (San Francisco Press, 1993), pp.64-69. Another version, 'Astronomical Engineers', was presented as a paper to SHOT, the Society for the History of Technology, at its Conference in Pittsburgh, October 2009.

1890. He was a talented inventor, being best known for devising the steam hammer which became an essential instrument of all forges and heavy machine shops in the second half of the nineteenth century. He was also a considerable draughtsman, a skill which he probably acquired as the son of Alexander Nasmyth, the Scottish portrait and landscape painter. And above all, for our present purposes, James Nasmyth had a life-long interest in astronomy, and was prepared to put all his skills at the disposal of this science. In the process, he made notable contributions to the design of telescopes, and to solar and lunar observations, which entitle him to recognition as an outstanding engineering astronomer.

Nasmyth had trained as an engineer in the workshops of the great mechanical engineer and machine-maker Henry Maudslay (1771-1831). Maudslay was impressed by the young man, just turned 20, who came seeking employment, having already designed and built steam engines and steam road vehicles, and took him on as a personal assistant. This appointment enabled Nasmyth to move around the workshops in Lambeth, to observe at first hand Maudslay's superb lathes and other metal-shaping machines, and to learn about his rational organisation and insistence on a very high level of accuracy in all the work produced. The young man was a quick learner, so that by the time Maudslay died in 1831 he had acquired sufficient experience and confidence to set up his own business, and in order to do this he shrewdly decided that his prospects as an engineer would best be served by a move to Manchester.

The town was thriving at this time as the heart of the expanding industry manufacturing cotton goods, creating a tremendous demand for machinery and for steam power, and its industrial vitality had been recently stimulated by the opening of the Liverpool & Manchester Railway – the first fully operational railway system in the world. Nasmyth made his home at Patricroft, on the edge of Manchester, where he established his famous factory, the Bridgewater Foundry. It was here that an astute French visitor, on being shown Nasmyth's notebooks by an assistant, was quick to appreciate the possibilities of the steam hammer which Nasmyth had sketched there for a challenging commission. This was a huge forging for the shaft of a paddle-wheel steam engine to equip I.K. Brunel's ship the s.s. *Great Britain*, which had then been put aside because Brunel had decided to adopt screw propulsion. The visitor had meanwhile returned to Le Creusot and built the first steam hammer. It did not take long for Nasmyth to assert his proprietorial rights in the invention, and to turn it to advantage. About the same time, in 1840, he married Anne Hartop, the daughter of a Barnsley ironworks manager. They had no family, but appear to have been a devoted partnership. Their house, 'Fireside', was alongside the Bridgewater Canal, and Nasmyth liked to describe his amusement at finding how his occasional sorties into the garden at dead of night wearing his night-shirt and carrying his telescope were sometimes observed by passing bargemen who promoted the belief that the house was haunted by a ghost seen flitting round the grounds 'with a coffin in its arms'.[3]

Manchester fulfilled Nasmyth's expectations, enabling him to find plenty of ready customers for his machine tools and to prosper in his business. He received help from members of his family and from people prepared to invest in his enterprise, which allowed

The garden of 'Fireside', Nasmyth's home at Patricroft, Manchester. (Autobiography, facing p.315)

him to expand rapidly even though Nasmyth gave little acknowledgment of this assistance in the pages of his *Autobiography*. It is possible that this reflects the emphasis of his editor on 'self-help', but it seems more likely that it is the result of Nasmyth's own view of himself as a self-made man because he expresses strongly the dogmatic individualism of the *laissez-faire* capitalist ethic which was the conventional wisdom amongst the aspiring middle classes of his day. It seems likely that Samuel Smiles received from him a well-structured body of notes that required very little editorial amendment, and that he was happy to accept this arrangement. So the image of Nasmyth as something of a loner in his business success is probably an accurate representation of his mind, and helps to explain his animus against the rising power of trade union organisations which, on his own admission, was one of the reasons for his decision to retire early from engineering manufacture. Having made a fortune from his machine tools and steam locomotives, and especially from the steam hammer and its various derivatives in pile-drivers and marine engines, Nasmyth thus put his wealth into long-term investments on which he and his wife lived comfortably thereafter in the large new house they acquired in Penshurst – even though he acknowledged the source of his wealth in the name of the house: 'Hammerfield'.[4]

In addition to his business concerns, the Manchester connection was important to Nasmyth because it brought him into contact with like-minded intellectuals and entrepreneurs who encouraged his astronomical interests through membership of the Manchester Literary and Philosophical Society and other similar bodies.[5] Scientific activities of many sorts enjoyed remarkable popularity in the city at this time, promoted by a plethora of organisations of which the MLPS was the oldest, having been established in 1781, and with figures such as John Dalton and Eaton Hodgkinson supporting it, as well as engineers such as William Fairbairn and James Nasmyth. Morrell and Thackray put it neatly: 'Science was indeed a dominant cultural mode in Manchester'[6], and Robert Kargon, in his study of this scientific activity in Manchester, places Fairbairn and Nasmyth in his second stage of scientific development, the 'devotees' stage, being men of substance who emphasised the need for professional commitment to science without themselves having had any scientific training.[7]

Nasmyth also made national friendships amongst scientists, through organisations such as the British Association for the Advancement of Science, which had been established in 1831 when the first of its series of annual meetings to stimulate public interest in science was held in York. He does not seem to have been a keen joiner of societies, as I have been unable to establish membership for him in the Royal Astronomical Society, the Institution of Civil Engineers, or even the Institution of Mechanical Engineers. But in 1842 he became involved in the local organisation of the BAAS for its meeting which was held in Manchester that year. Sir John Herschel (1792-1871), the son of William Herschel, and Warren de la Rue (1815-1889), were both prominent astronomical members of the Association, and Nasmyth became a close friend of both of them. Both had promoted the development of photography, which was still in its infancy in the mid nineteenth century. Herschel had pioneered the study of light-sensitive materials which was then pursued in France by Niépce and Daguerre, and had been one of a group of scientists and engineers who had encouraged Fox Talbot to develop his negative images, which could then be used for any number of positive prints.[8] Astronomers were anxious to take advantage of this new technology, but were hampered by the low luminosity of most of their evidence at a time when photographic plates still lacked sensitivity and required long exposures. At the other extreme level of luminosity, solar photography required heavy filters which obscured delicate detail.

Unlike engineering, which had become a well-organised and respected profession in the nineteenth century, most scientific activity remained an occupation of amateurs at this time. There were a few salaried officers in the employment of bodies such as the Royal Institution who were able to make substantial contributions to the advance of scientific knowledge – the understanding of the natural world – but for the most part science remained the preserve of gentlemen of an assured livelihood and plenty of leisure, which they were prepared to devote to the investigation of natural phenomena. This has been well demonstrated by Morrell and Thackray in their excellent study of *Gentlemen of Science*, discussing the early history of the British Association for the Advancement of Science, who show how the remarkably vigorous scientific activity in Britain in the middle decades of the nineteenth century was led by a collection of such devoted amateurs.[9]

Astronomy, in particular, depended heavily upon the diligent observations of people for whom the labour provided no financial rewards. Salaried officers were rare, although an Astronomer Royal had been appointed in 1675 in the person of John Flamsteed (1646-1719), and his successor Edmond Halley (1656-1742), appointed in 1721, accurately calculated solar eclipses and the return of the comet named after him. But most of the significant developments were made by amateurs such as William Herschel, who had begun his career as a musician and music teacher in Bath. Herschel's inspired perception that the Milky Way might define the limits of our own galaxy, with other galaxies existing as faint nebulae beyond it, was confirmed by the observations of Lord Rosse, an aristocratic amateur, that many such galaxies existed. This was made possible because Herschel's superb but essentially home-made telescopes were superseded by the mighty reflecting telescope erected in 1845 by Lord Rosse on his estate at Parsonstown, and by the deployment of a host of new instruments. The reliance upon skilful measurements made by hand and eye was converted in the nineteenth century to dependence upon the precision of photography and the flood of new information from the spectrometer, whereby it became possible for the first time to discover the chemical composition of the sun and other stars.[10] A greatly expanded view of the universe meant that interest shifted from close study of the sun and moon to a concern for the unfathomable mysteries of deep space. These developments coincided with the onset of rapid industrialisation in the West and the corresponding globalisation of economic and political structures, all processes in which the engineering profession played an important part. As a typical gentleman astronomer of his generation, James Nasmyth made some strikingly original contributions both to instrumentation and to observations which won the praise of fellow astronomers.

In his *Autobiography* Nasmyth tells how Maudslay, towards the end of his life, developed a keen interest in astronomy, and how he – Nasmyth – was able to instruct his master in some of the principles of grinding a telescopic mirror from speculum metal, a process which he had developed experimentally in his youth. In the course of many attempts to make his own telescopic mirrors, he had devised an ingenious technique for casting the alloy (32 parts copper, 15 parts of grain tin, and one part of white arsenic) in an open cast-iron mould rather than a closed sand mould, which greatly reduced the propensity of the disc of speculum metal to become extremely brittle. This facilitated the grinding of the disc to give a brilliantly reflective mirror of eight inches diameter or more.[11] Unfortunately, Maudslay died before Nasmyth could complete his projected telescope, but he sustained his own interest in the subject and went on to design and build many excellent astronomical instruments. In particular, he developed a 'trunnion vision' telescope, in which the light from the secondary mirror was extracted through the hollow trunnions of the telescope mounting, providing maximum convenience for the observer, who could sit comfortably at the eye-piece on a platform still known on many modern telescopes as the 'Nasmyth platform'. He demonstrated this device very effectively through a lecture delivered by his friend Edward Cooper to the Royal Institution in 1849.[12] He also devised a mounting for the telescope on a railway turntable, giving excellent manoeuvrability to the instrument.

Nasmyth used a succession of telescopes made in this way to make a close study of the sun. There was a total eclipse of the sun in 1860, visible from northern Spain, and the railway engineer Charles Blacker Vignoles (1793-1875) – another engineer with astronomical interests – who was working on the Bilbao and Tudela Railway at the time,

Nasmyth at the 'trunnion vision' of his 20-inch diameter telescope, on a turntable. (Autobiography, p.338).

was able to use his position as Chief Engineer to promote a scientific expedition to Spain in order to observe the eclipse on 18 July. He solicited naval support to make a transport vessel available to carry the party, and he put his staff at the service of the astronomers under the Astronomer Royal, Professor Sir George Airy (1801-1892).[13] The phenomenon of a total solar eclipse is a rare event, and as it only lasts for a few minutes in which the body of the moon totally covers the disc of the sun, few astronomers are able to take advantage of such events to witness the extraordinary features which it displays: the sky goes dark, the stars appear, the moon is seen surrounded by a pearly halo, and plumes of light appear round its edges. Early observers considered that the pearly light – the 'corona' – was the rarefied atmosphere of the moon, and that the bursts of light at the edge of the moon – the 'prominences' – represented volcanic activity on the lunar surface.

Vignoles also made arrangements for Warren de la Rue to accompany the expedition in order to attempt to take the first photographs of a total eclipse. These photographs were subsequently reproduced in the *Philosophical Transactions* for 1862 as an appendix to de la Rue's massive Bakerian Lecture to the Royal Society on the subject of the 1860 eclipse. The 'reproductions' were in fact touched-up drawings, although it is possible that the original plates were deposited with the Royal Astronomical Society. Amongst other significant discoveries, they established that the 'prominences' observed were a solar phenomenon, rather than a trick of the light on the edge of the moon. The corona was too faint to register on the available plates, but the observations suggested that this also was definitely a solar feature rather than a lunar atmosphere. This was the first occasion on which it had been possible to study photographically the course of a solar eclipse and these interpretations of the prominences and the corona both represented important advances in astronomical understanding. De la Rue's paper on the eclipse of 1860 referred briefly to some equipment for observing solar prominences devised by James Nasmyth.[14] It consisted of 'a cylindrical box, blackened inside' which admitted the image of the sun through an aperture, reducing the intensity of the light and allowing background light to be registered when the sun was eclipsed. The device failed to work satisfactorily, but it probably helped to record several solar prominences even though the fainter pearly light of the solar corona did not appear on the photographs.[15]

The sun was one of Nasmyth's long-standing subjects of scientific investigation, and he was the first astronomer to draw attention to the mottled nature of the solar surface, which appeared to him to be composed of minute willow-leaf grains of luminous material. This discovery, puzzling at a time before the understanding of nuclear fusion and violent electro-magnetic activity had been acquired, and when the common mindset saw the sun as a gigantic coal-fire consuming its own material, was made by Nasymth as a result of careful observations under favourable conditions, and expressed in beautifully executed illustrations. Although at first reluctant to accept this discovery, Herschel and other astronomers were persuaded by the excellence of Nasmyth's observations. Herschel was not the most effusive of men, but in 1861 he wrote to Nasmyth:

Group of sun spots as observed by Nasmyth in 1864, showing the 'willow leaf' granulation of the solar surface. (Autobiography, facing p.370)

> I suppose there can be no doubt as to the reality of the willow-leaved flakes, and in that case they certainly are the most marvellous phenomena that have yet turned up – I had almost said in all Nature – certainly in Astronomy.[16]

He subsequently incorporated a report on Nasmyth's observations in a new edition of his classic study, *Outlines of Astronomy*, and wrote warmly of their significance in his *Popular Lectures on Scientific Subjects* in 1884.[17] Confirmation of Nasmyth's observations was received from the Royal Observatory, Greenwich, and from other astronomers. De la Rue wrote to Nasmyth:

> I like good honest doubting. Before I had seen with my own eyes your willow leaves, I doubted their real existence … But when I actually saw them for the first time, I could not restrain the exclamation, 'Why, here are Nasmyth's willow leaves!'[18]

Without any doubt, Nasmyth had made a significant contribution to solar astronomy.

Nasmyth was amongst the first astronomers to apply his ingenuity to the resolution of the problems of astronomical photography. Apart from his assistance of de la Rue, his distinctive contribution was that of transferring information from the meticulous drawings and models which were a family speciality onto photographic plates. He devoted time and attention to developing this technique in the 1840s even though at that time he still remained extremely busy in pursuing his engineering enterprise. He won the praise of many visitors to the BAAS meeting in Edinburgh in 1850 and then to the Great

Exhibition of 1851 for his display of pictures of the lunar surface. He won a Prize Medal for his lunar drawings at the latter, and attracted the special notice of the Prince Consort who then arranged for a demonstration to Queen Victoria who 'took a deep interest in the subject, and was most earnest in her inquiries'.[19]

Despite his attention to the sun, Nasmyth's first love, amongst astronomical observations, was the study of the lunar surface, and it was here that his photographic technique was most fully exploited. When he came to write a substantial book on *The Moon: considered as a planet, a world, and a satellite* in collaboration with John Carpenter (who worked at the Royal Observatory, Greenwich), published in 1874, he was able to make good use of his acquired skill in combining photography with careful observation and precise drawing.[20] The book expounded a volcanic explanation of the origin of the lunar craters, worked out in detail and expressed very impressively in his abundant illustrations, and was praised by Sir John Herschel and other professional astronomers. But there were difficulties about Nasmyth's theory which were expressed by Sir Robert Ball, in his book *The Story of the Heavens*:

> The most probable views on the subject [the formation of the lunar craters] seem to be those which have been set forth by Mr Nasmyth, though it must be admitted that his doctrines are by no means free from difficulty.[21]

Ball was particularly concerned about the great size of some of the lunar craters, which made it appear unlikely that volcanic activity alone could have been the cause of their formation.

Central to Nasmyth's explanations were a number of homely images such as the wrinkling of an apple – or even a human hand and face – as they age, the behaviour of bubbles, and the cracking of glass by heat to create a radial pattern. He was quick to notice correspondences and to exploit their significance, but his engineering imagination let him down in this instance, as astronomers now accept as a matter of course that the pocked appearance of the lunar surface and the radial diffusion of ejecta from craters like Tycho and Copernicus are not primarily of a volcanic nature but rather the result of the prolonged bombardment by debris in the early formation of the solar system.

Nevertheless, Nasmyth's explanations were supported by excellent illustrations, depicting lunar features in great detail. These were made from drawings built up over many successive observations: the drawings were then converted into plaster or *papier-mâché* models which were subjected to an oblique light to create the desired shadow effects, and the images were then photographed to create some hauntingly authentic pictures. Even though the volcanic theory of the lunar craters has subsequently been abandoned, Nasmyth's accurately detailed illustrations of the moon – together with those he made of the surface of the sun to demonstrate its mottled appearance – were frequently reproduced, and fully entitle him to be taken seriously in the history of astronomy. So highly regarded, indeed, were Nasmyth's illustrations that it seems as if many astronomers were prepared to accept them as real photographs, although this led to an expectation of lunar mountains being more rugged than they were found to be by the first Apollo spacemen. One recent

(above) Section of the lunar surface as modelled by Nasmyth. (Autobiography, p.322)
(below) Model of a lunar crater by Nasmyth, showing the purported volcanic origins. (Autobiography, p.318)

commentator has suggested that some well-established astronomers seemed to want to believe that these brilliant evocations were genuine photographs, even though they knew they were not so.[22]

Nasmyth thus deserves credit for several achievements as an astronomer. His contributions to the design of telescopes were important, and his impressive observations of the sun and moon won the respect of the astronomical fraternity. He encouraged experiments with photography, but used this new technique mainly in a secondary role, to create images from his elaborate lunar models, and he was not actively involved in the exciting contemporary developments in spectroscopy. The fact that his analysis of the process of lunar crater-formation has been rejected does not detract from the stimulus to research which it provided. Likewise, Nasmyth's perception of the extraordinary activity in the solar photosphere – the luminous outer surface of the sun – gave a considerable impetus to the investigation of sun-spots, solar prominences, and the solar corona. In particular, it led Nasmyth to some surprisingly modern speculations about the nature of light and solar energy, such as his judgment that the main source of solar light 'appears to result from an action induced on the *exterior surface* of the solar sphere', and that it probably results from the action of the sun in consuming material from inter-stellar space. He suggests, further, that the uneven distribution of such material could account for fluctuations in solar energy and thus produce Ice Ages and associated phenomena.[23]

Nasmyth generally gave little attention to the more far-reaching aspects of astronomical enquiry, such as the spiral configuration of some of the *nebulae* – the wispy clouds that perplexed so many of his astronomical colleagues – although when he did think about the subject he did so in characteristically matter-of-fact terms such as likening it to the way in which water goes down a plug-hole. He speculated that:

> The first moment of the existence of such a nebulous mass would be inaugurated by the election of a centre of gravity, and, instantly after, every particle throughout the entire mass of such nebulae would tend to and converge towards that centre of gravity … inducing thereby a twisting action because of the random dispersal of its components, from which rotation would ensue and produce the distinctive spiral patterns of the galaxies.[24]

In most respects, Nasmyth's astronomical interests were traditional rather than innovative, but these perceptions show that he was not averse to speculation, as his images of ageing skin and cracking glass in his discussion of lunar features demonstrate, even though these images tended to be homely rather than conceptual, avoiding any serious challenge to cosmic orthodoxy. There can be no doubt, however, that Nasmyth was able to apply his engineering skills to brilliant effect in his astronomical observations and speculations. Thus, while it would not be justified to generalise from this experience of a single engineer that there is an essential connection between such practical skills and astronomical competence, it is certain that James Nasmyth contributed to significant advances in astronomy, as well as being the engineer and industrialist who made a fortune out of the steam hammer.

Notes

1. See, for example, A.E. Musson, 'James Nasmyth and the early growth of mechanical engineering' in *Economic History Review* (2nd series, vol.10, 1957-58); H.W. Dickinson, 'James Nasmyth as a toolmaker' in *The Engineer* (23 May 1941); and J.A. Cantrell, *James Nasmyth and the Bridgewater Foundry* (Manchester: Chetham Society, 1984). I was able to include a paragraph on Nasmyth's astronomical interests in the recent new edition of the *Oxford Dictionary of National Biography*, under 'Nasmyth, James Hall'. The fullest account of Nasmyth's astronomical investigations is that compiled from his own notes by Samuel Smiles (ed.), *James Nasmyth, Engineer: An Autobiography* (London: John Murray, 1885).
2. L.T.C. Rolt, *Tools for the Job: A Short History of Machine Tools*, (London: Batsford, 1965): the main section on Nasmyth is pp.108-113; *Victorian Engineering* (London: Allen Lane, 1970): see chapter 5, 'The Workshop of the World', pp.116-147.
3. *Ibid.*, pp.315-6, gives this anecdote, with a drawing by Nasmyth (p.314) of his garden at Patricroft: another drawing on p.338 illustrates the 'trunnion vision'.
4. Nasmyth's strong feelings about trade unions are expressed in Smiles, *op.cit.*, pp. 213-217: 'Hammerfield' is illustrated in a plate on p.360.
5. Robert H. Kargon, *Science in Victorian Manchester; enterprise and expertise* (Manchester: Manchester University Press, 1977) has a good account of this activity: he described nineteenth-century Manchester as 'a splendid, shining world-center for scientific research and teaching' (p.1).
6. Jack Morrell and Arnold Thackray, *Gentlemen of Science: Early Years of the British Association for the Advancement of Science* (Oxford: Clarendon Press, 1981), p.397.
7. Kargon, *op.cit.*: this was preceded by an 'amateur' stage, from 1781 to the 1830s; and was followed by a stage in which professional 'civic' scientists took the lead; and a fourth stage of 'academic' science, beginning with the foundation of Owens College in 1851; developing into a fifth stage of 'university' science with the establishment of Victoria University in 1880.
8. William Henry Fox Talbot (1800-1877), pioneer photographer, published *The Pencil of Nature* in 1844, the first book to be illustrated entirely by photographs. Fox Talbot took striking images of the Hungerford Suspension Bridge in London, and the s.s. *Great Britain* in Bristol Harbour in 1844. But early attempts at astronomical photography were disappointing.
9. Morrell and Thackray, *op.cit.*
10. Gustave Kirchhoff discovered in 1859 that every pure substance has its own distinctive spectrum, and the recognition that such signals could be identified in the spectra of stars meant that it became possible to define their composition and to work out the stages of stellar evolution.
11. Smiles, *op.cit.*, pp.169-70.
12. Smiles, *op.cit.*, p.340.
13. Olinthus J. Vignoles, *Life of Charles Blacker Vignoles* (London: Longman, 1889), p.357.
14. Royal Society, *Philosophical Transactions*, (1862), p.406.
15. Royal Society, *Philosophical Transactions* for De la Rue's Bakerian Lecture (1862), pp.333-416. I am grateful to Nick von Behr and Emma Lambourne of the Royal Society Library for their help in pursuing this reference.
16. For Nasmyth's solar observations, see Smiles, *op.cit.*, pp.370-374, and plate opposite p.370. The letter from Herschel to Nasmyth was dated 21 May 1861 and is quoted in Smiles, *op.cit.*, pp.370-371.
17. Sir John F.W. Herschel, *Popular Lectures on Scientific Subjects* (London: W.H. Allen, 1884), p.83.
18. Warren de la Rue to Nasmyth, March 1864, quoted in Smiles, *op.cit.*, p. 373. Nasmyth's initial report was made in *Memoirs of the Literary and Philosophical Society of Manchester*, 3rd series, vol.1, p.407 and appears as a footnote on p.370 of the *Autobiography*.
19. Smiles, *op.cit.*, p.333; Nasmyth quotes the Queen's *Diary*, on the same page (referring to Sir Theodore Martin, *Life of the Prince Consort*, vol.2, p.398).
20. James Nasmyth and James Carpenter, *The Moon: considered as a planet, a world and a satellite* (London: John Murray, 1874). See also Smiles, *op.cit.*, pp.374-382, including plates on pp.376 and 377 which offer an explanation of the radial lines on the surface of the moon. It is not possible to determine exactly Carpenter's contribution to the book, but it seems likely that Nasmyth provided the bulk of the text, and certainly his illustrations dominate the work.
21. Sir Robert Stawell Ball, *The Story of the Heavens* (London: Cassell, 1886), p.72 (1891 edition).
22. Frances Robertson, 'Science and Fiction: James Nasmyth's Photographic Images of the Moon', in *Victorian Studies*, 48.4 (2006), pp.595-623.
23. Smiles, *op.cit.*, pp.341-343, reporting a paper sent to the Royal Astronomical Society in May 1851.
24. Smiles, *op.cit.*, pp.345-347: from a report sent by Nasmyth to the Royal Astronomical Society in 1855.

6: Engineering Education in the Age of Microelectronics

Robin Morris

When compared with the older, well-established, branches of engineering, microelectronics engineering has certain features which set it somewhat apart from its predecessors. For example, it crosses traditional academic boundaries to a much greater extent, and it is highly theoretically-based, investigating the properties of materials using theoretical models operating at the electron level. The explanation of the behaviour of such particles as electrons rests upon theoretical concepts which form the basis of quantum physics. The influence of contemporary physical science therefore looms very large in the world of those engaged with the field of microelectronics.

Those principally involved in the development and production of microelectronic devices are usually called semiconductor engineers. Their work encompasses not only a theoretical understanding of the properties of a wide range of materials and processes, but also the ability to use this knowledge in order to manufacture a wide variety of solid-state electronic components and integrated circuitry. To do this, a wide diversity of processes and techniques needs to be mastered, including, for example, metallic alloying, photographic etching, and the diffusion of gases. These activities demand at the very least some acquaintance with chemistry and metallurgy, to a greater degree than most other branches of engineering. The practising semiconductor engineer must therefore acquire a wide range of practical experience, and be adept at working to extremely high levels of accuracy. This situation suggests that the training of semiconductor engineers might well need to follow a somewhat different path from that currently offered within traditional 'common core' engineering degree courses.

The manufacture of solid-state diodes began in Britain before the Second World War, when production expanded greatly. But the semiconductor industry only really began with the manufacture of transistors, this event dating in Britain from the early to mid-1950s. It was an entirely new venture, and at the time very few people even knew what a transistor was, let alone how to manufacture one. (It was even claimed that the management of British Telecommunications Research were then prepared to recruit anyone who had even heard of the word transistor!) Faced with this situation, the learning curve was steep, as electronic engineers, physicists, chemists and metallurgists were hastily recruited to form teams within the existing electronics manufacturing companies. This was a good time to enter such an exciting new field, and I was only one of many others attracted to join them. The few existing 'experts' were as rare as gold dust, whereas most of the rest were endeavouring to learn off each other.

Consequently, after a decade or so, a body of semiconductor engineers existed which had been recruited from a wide range of backgrounds and disciplines. They retained, to a greater or lesser extent, something of the characteristics of their former training, including their approach to the solution of problems. At this time there were of course no formal academic courses specifically devoted to semiconductor engineering, and so lecturers in the subject just had to 'pick it up' like the rest of us, perhaps by visiting the United States, which had already established a firm technological lead in the field. This somewhat unusual situation offered a window of opportunity for people like myself, coming from a rather unorthodox background, to place our feet on the bottom rung of what turned out to be an extremely helpful ladder.

Today, the engineering industry is in a state of constant change and so adaptability is all-important. The days are long gone when someone could complete a seven-year apprenticeship, find a niche somewhere in a factory, and then plod on steadily doing much the same job until retirement. Driven by the rapidly changing electronics industry, and in particular by microelectronics technology, all other branches of the profession are now obliged to come to terms (whether they like it or not!) with increasingly complex electronic automated processes in one form or another. As a result, traditional distinctions between the professional engineering societies are now being steadily eroded as each is subjected increasingly to the inroads of electronics-driven technology.

The growing complexity of microelectronics technology has also influenced the way in which education and training are conducted within the universities, with the organisation of practical work in academic research presenting an increasing problem. In particular, because of the complexity and expense of semiconductor fabrication equipment, any leading-edge research and development involving production processes becomes much more difficult to arrange. This can only result in moving such leading-edge research projects away from university departments towards industrial research establishments, where they suffer the disadvantage of becoming increasingly directed into short-term commercial interests. Unless adequately funded, university research departments can only trail in the wake of research programmes devised and directed by the large semiconductor manufacturers, or alternatively restrict postgraduate research to specialised 'niche' areas where the constantly rising cost of capital equipment is not beyond the means of departmental budgets.

With ever-increasing technological complexity the problem of training specialised research workers will only intensify. One consequence of this is that the technological gap already existing between the training requirements of the average electronics undergraduate and that of the highly specialised research worker will further widen. So far, this problem has been met either by including a particular subsidiary specialism within a general engineering degree course, or – much more effectively – through specific postgraduate study. Work at this level, however, is very intellectually demanding, so that it is essential to attract the very best students available and it is worth considering whether there might be some better way of training the high-flying, highly specialised, engineer. Such individuals might well benefit from a different sort of initial training, involving, for example, a fast-track and in-depth course of studies right from the start. This arrangement would ideally involve setting up a well-funded, high-status institution, specifically dedicated to training engineers and scientists capable of holding their own in competition with those now emerging from the highest centres of excellence overseas. Such an institution would aim to attract staff of outstanding calibre, and so create a research 'skill-cluster' capable of holding its own at the highest international level. It would be extremely important to ensure that such an institution was backed by adequate public funding, and therefore kept free from the influence of vested industrial interests.

As a pathway to entering leading-edge microelectronics research, the broadly-based, 'one-course-fits-all' electronics engineering degree – which may include some element of specialist study introduced at some stage during the course – is inadequate. Nevertheless, it seems difficult to question the need to continue to run general courses in the various branches of engineering in order to provide an efficient base for current industrial demand. Most engineers would probably agree that such courses should be flexible enough to cope with future unknown developments, as well as providing a sound base for future study. Because the engineering profession has for some time being undergoing a form of technological convergence, largely under the influence of the strongly scientifically-based electronics industry, it seems necessary to establish a common core of instruction for the profession as a whole, and furthermore to ensure that this common core is strongly scientifically based. The primary aim of such a policy should be to produce scientifically trained engineers, well-grounded in basic principles, who are both versatile and adaptable.

Following its creation in 1965 the Council of Engineering Institutions (CEI) recognised the need to move in this direction and argued for a common-core engineering syllabus upon which the various engineering institutions could then superimpose their individual specialist requirements. A strong motive behind this policy was to strengthen the unity of the profession and to raise its status as a whole. It was arranged, therefore, that on successful completion of a course of approved academic level, followed by a period of appropriate practical experience, the title C.Eng. (Chartered Engineer) would be awarded by the Engineering Council.

Although it is certainly necessary to run common-core engineering courses it is questionable whether they form the right basis for the highly complex and specialised work which will undoubtedly take place within the laboratories of the leading industrial nations during future years. If sufficient engineers of the highest calibre are to be trained, capable of spending many years working within an increasingly complex and rapidly changing industry, whilst at the same time maintaining a high degree of specialisation, then we are faced with the probability that, by pursuing a core engineering course, both time and enthusiasm will have been lost by the highly important minority of 'high-flyers' needed for such advanced work.

As matters stand at present, no attractive route exists within engineering education in Britain which directly addresses the problem of recruiting and training such 'high-flyers'. It is well known that many ambitious talented individuals living in Britain today, just as in the past, do not choose to enter the engineering profession, preferring instead to join occupations enjoying greater social prestige and financial reward. Perhaps a litmus test in this regard might be to ask how many of those educated at the major public schools would consider choosing engineering as a career, in preference to (say) the legal profession or medicine. It is therefore important to address the questions of both status and reward. One solution would be to offer alongside general engineering degree courses (now based much on the lines of those agreed by the CEI) a 'fast-track' programme of study for the very bright students. Although fast-track in tempo, such a programme would of course need to last substantially longer than a general engineering degree. A key element would be the inclusion of practical work at a high level, organised in well-funded laboratories. This should certainly not preclude industrial links, although remaining strictly independent regarding funding and consequently policy.

A problem to be faced squarely is that in order to attract sufficient numbers of highly talented individuals into 'fast-track' engineering courses there must be parity of esteem when compared with older professions including barristers and senior medical staff. To achieve this state of affairs is obviously no mean task, but perhaps the most important single step would be to offer such 'fast-track' courses within some prestigious national institution, equivalent in public esteem to the Massachusetts Institute of Technology in America or the French École Polytechnique. Such advanced engineering courses would need to carry a correspondingly distinctive award, equivalent to the French Diplome d'Ingénieur or the German Dipl.Ing. Both these qualifications, most importantly, carry an official status.

An important spin-off from such a decision would be to raise the status of engineering as a whole. This matter of status has been a matter of perennial complaint among engineers in all disciplines and at all levels. It is perhaps significant that America, France and Germany all possess prestigious engineering institutions and these countries all publicly assign a far higher status to the engineering profession than does Britain. I have also gained the impression, in recent conversation with Chinese engineers, that engineering in China is highly regarded. A further attraction would be that the achievement of higher status for the profession would of course bring with it the prospect of increased financial reward. A possible objection might be that some individual engineering institutions, by the nature of their work, might find high-level theoretical studies of the nature envisaged inappropriate for their members. Although regrettable, this would, I believe, be a small price to pay in order to build a world-class, scientifically-based engineering profession capable of holding its own upon the future international scene.

A highly important consequence of the failure of successive governments to improve the status of the engineering profession is that it has so far been difficult if not impossible for engineers and technical personnel, either in industry or in government research establishments, to gain face-to-face access with government ministers and so directly influence public policy. This situation appears to be largely due to the dismissive attitude of the predominantly non-scientifically trained senior civil servants who control and to a great extent monopolise such access. A major difficulty would therefore be to persuade the well-entrenched, classics-educated senior government advisers to concede an influential role to the engineering profession or reward its members by promotions to positions of executive influence. Positive measures by government to incorporate an officially recognised engineering elite into the system, including access up to cabinet level, would undoubtedly help to change attitudes in this regard. It is well known that the situation in Britain contrasts greatly with that in countries such as France, where engineers often hold positions in the highest levels of administration and are consequently capable of understanding and evaluating technical advice. If the engineering profession cannot 'punch its weight' at a nationally equivalent level, and achieve a measure of influence in government circles in proportion to the importance of the industry, then our decision makers will continue to be denied the unrestricted opinion of technical experts when framing public policy.

The peculiarly British attitude of disparagement towards engineering is not confined to government bodies, but is also met with on the boards of directors of many of the largest companies, where accountants seem usually to predominate. If public perceptions of engineering are to change for the better, acceptance of parity of esteem both by industry and by public bodies is essential. The fact that promotion of engineers to positions of real influence occurs relatively rarely in Britain, unlike the situation in other technologically advanced countries, can hardly fail to discourage the highly able and ambitious student from either entering or staying within the engineering profession. Instead, many able students, when realising the extent of the problem, move during their studies to other activities offering higher social esteem and ultimately better career prospects.

It is generally recognised that the establishment of the CEI in 1965, set up in order to speak with one voice on behalf of the various professional institutions, was a welcome, albeit somewhat belated, move in the right direction. Also welcome was the well-meaning attempt by the Council to improve both the status and unity of the profession as a whole through the establishment of the C.Eng. Diploma. However, when viewed with hindsight, these measures do not seem to have been thoroughgoing enough to deal successfully with the problems they were intended to solve. What was really needed in order to make any difference in status was the creation of a high-level institution for training engineers which would be capable of claiming parity of esteem with its foreign counterparts, and awarding qualifications recognised as being of an equal standard. For whatever reason, this opportunity was lost.

It is not possible here to attempt a precise definition of what such an institution would involve, but some personal points are appropriate. Firstly, entry into a university engineering undergraduate course should require a good solid grounding in pure and applied maths, physics and chemistry at GCE 'A' Level, together with at least a good pass in GCSE English. In addition, some acquaintance with biology, and in particular cell structure, might prove to be of future value. This would then provide a sound base for future studies, whatever branch of engineering was chosen. Specialised courses at school, such as electronics, although no doubt highly interesting to the class teacher, seem to be of little value and do hardly more than detract from obtaining a good grounding in the subjects already mentioned.

To complete a 'general course' engineering degree to honours level, three years of study should be the bare minimum. Some universities have already moved to four-year courses, and this is to be applauded. The university mathematics syllabus should obviously have a strong engineering slant, and certainly include some statistical analysis and reliability theory. Useful subsidiary courses of study would be metallurgy and further chemistry. If additional subsidiary subjects are to be included, such as management theory, then the three-year period should definitely be extended. Emphasis should always be placed upon illustrating basic principles and fostering a scientific approach to problem-solving rather than concentrating unduly upon ephemeral technological detail. Practical work should always be seen by the student to be meaningful, and involve some specific individual challenge, such as building and getting a particular piece of equipment to function successfully. Early success at practical problem-solving is important to build confidence. Ideally, an extension of several months beyond the three-year period of study, devoted entirely to practical laboratory work, would be well worth the time spent.

It is always important in designing training programmes for engineering courses to find the right balance between practice and theory. Ideally, the aim should be to achieve practical proficiency, resting upon a sound theoretical understanding. Generally speaking, however, British engineering practice has in the past too often stressed a predominantly empirical approach to problem-solving, its practitioners sometimes even openly despising scientific methods of analysis. Sir Eric Ashby drew attention to this lack of a deliberate and systematic link between theoretical science and technological invention in British practice.[1]

American and Continental engineers have, in the main, avoided this earlier acquired tendency, and done so much to their advantage. Henry Malden, in a lecture as early as 1845, emphasised the importance of the sciences in sharpening habits of observation, accuracy, and logical thought, and although he was referring to the physical sciences his insights apply equally to engineering.[2] It should certainly be the aim of every branch of engineering education to instil a scientifically-based outlook. The fact that it is necessary to preach the same sermon today, over 150 years later, suggests that the ghost of British rule-of-thumb empiricism may still linger, dimly hovering over what remains of the British engineering industry.

The British engineering industry is today at the crossroads. We can no longer compete with low-cost economies within the field of mass production and so we must strive to maintain a place amongst technological world leaders in research as well as manufacturing the most advanced and complex equipment. This can only be done if adequately funded skill-clusters of sufficient size exist. To attempt to establish such skill-clusters on a tight budget invites failure, since past experience points to the fact that our overseas competitors will certainly not subject themselves to undue financial constraints in the field of high technology. It is essential, therefore, that the very best engineering training should be provided, and should be made a national priority. Short-term half measures and *ad hoc* policies, traditionally favoured by successive governments, have failed in the past and are hardly likely to succeed in the future. Microelectronics is central to the future of engineering in Britain. Perhaps it may now be beyond the means of Britain alone to fund such projects, although the resources of the European Union would certainly be adequate. Failure to keep abreast of other powers within the field of electronics and in particular microelectronic engineering can only lead to further economic and political decline.

Notes

1 Sir Eric Ashby, *Technology and the Academics* (London: Macmillan, 1958)
2 Henry Malden, 'On the Introduction of the Natural Sciences into Education', lecture at University College, London, 1845.

7: Testing Times: Aerospace and Historic Engines

Peter Stokes

I grew up in South London with practical interests, being good with my hands, a Meccano enthusiast, a 'cub' and then a Boy Scout. My interest in history and technology was fired by visits to our local Horniman Museum in Forest Hill, and to the Science Museum in South Kensington. My mother encouraged me further with steamer trips on the Thames, and visits to Croydon Airport with its Imperial Airways Handley Page 'Heracles', Lufthansa Junkers 52s, Swissair DC 2s, and exotic French airliners. Mum reminded me that she had been there with Dad on their motor-cycle combination to greet the arrival of Charles Lindbergh from Paris in 1927, following his Atlantic crossing. Our visit is fixed in my mind as being 1938 by the memory of seeing two RAF 'Wellesley' bombers on return from their long-distance, record-breaking flight to Arabia. Evacuation to Surrey early in the Second World War brought sights of RAF fighter defence, and then with the return home, the Luftwaffe in offence, and later the flying bomb and rocket attacks while I was studying at Technical School and its Air Training Squadron. It was all a useful introduction to the modern world and what was to be my industrial experience of power in aviation and aerospace.

In the mid-1940s I undertook my engineering apprenticeship with De Havilland Aircraft Company, while attending South East London Technical College to get my Higher National Certificate in Mechanical Engineering, as endorsed by the Institute of Marine Engineering. My thermodynamic studies had included laboratory experience in engine testing with steam, petrol and diesel engines. Achieving graduate status with the Institution of Mechanical Engineers and the Royal Aeronautical Society, I gained employment with the recently created De Havilland Engine Company in the drawing office of the Test Facility Department as a Draughtsman and Development Engineer. The Department provided a microcosm of design, development, manufacture, sub-contract and service: a highly motivating field in which to be employed, whilst enjoying the overall support structure of the parent company.

The Engine Company, in its emergence from the DH Aircraft Co, was very strong in innovation from its pre-war distinction with 'Gipsy' piston engines in record breaking and overall performance. A new range of piston engines had been designed for the post-war market, and an outstanding breakthrough had been made into jet propulsion with the 'Goblin' military engine, which would be followed by the more powerful 'Ghost' in both military and civil fields, most memorably in the first jet-propelled civil airliner, the 'Comet'. The company would continue to innovate in supersonic jet propulsion, rocket engines and helicopter turbo-shaft applications.

Piston engine test requirements are normally met in development and production by connection to hydraulic dynamometers to measure torque at various speeds, and hence calculate brake power. The company of Heenan & Froude had established a wide range of products in Britain, mainly water hydraulic, having a cubic power speed response curve compatible with a 'propellor' law. The torque reaction at speed was determined by a variable vortex 'gate' placed in obstruction between a driven rotor and a stator which were bladed in the manner of a fluid flywheel transmission. This arrangement was the core element of a 'test bed', complemented by air intake and exhaust arrangements, coolant circulation, and fuel and oil circuits with metering and instrumentation for all parameters.

The succession of Air Ministry to Ministry of Supply, and then Defence, had established a panoply of test routines for military requirements, and these were generally replicated across the international and civil fields. Ground-level testing was the standard procedure, with all performance criteria set to international standard atmosphere for density and temperature. Altitude testing is simulated by drawing atmosphere intake from a sealed box at a measured sub-atmospheric level with a throttled entry.

Jet Propulsion

With the advent of jet propulsion, a new field of test proving was introduced. De Havilland was one of the first in the field in both development and production. Propulsion thrust was initially measured by a mounting suspended on swinging links as a parallelogram, and later by a test trolley mounted on rollers and then on short flexural straps fore and aft, with all connections, fluid, electrics, and instrumentation made by radially disposed pipes. High accuracy was achieved with minimum hysteresis. Dead weight calibration arrangements were built into the test bed design.

The major rotative components, compressor and turbine rotors, required overspeed testing both in development and in production. This was undertaken in vacuum chambers within armour-plated ring shrouds and generally disposed with the axis vertical, as suspended from drive-geared electric motors, or by air or steam turbines above the chamber. The most critical component in gas turbine design is probably the compressor, with its high power requirement, exceeding the energy output of the engine. All designing companies, as supported by national institutions, had to put capital investment into compressor test plant. The power for these was variously provided by adaptation of power stations, electric motors working off demand from the national grid, and naval marine power plant brought ashore for this purpose.

The De Havilland approach in its Halford Laboratory at Hatfield was pioneering in harnessing the four 'Ghost' engine power plant of a complete 'Comet' aircraft to the job. This was an early example of the use of aero gas turbines for ground power requirements. The engine design organisation, with the stress office and performance departments, provided the design in cast-iron, water-cooled casings. Each pair of turbines produced 9,000 hp at 6,500 rpm, matched to a compressor test requirement for the new 'Gyron' large engine for supersonic flight. Our test facility department produced the overall installation with differing change-speed gearboxes at each end. In development we had to substitute fabricated for the cast-iron cases, with sacrificial

De Havilland's Halford Laboratory and test bed centre at Hatfield. (reproduced by courtesy of Rolls-Royce Heritage Trust)

anodes in the water-cooling chambers. We had whirling problems with the gear-toothed coupling high-speed output shaft and added a supplementary oil-pumped flow for the high-speed pinion plain bearing, countering instability. When proven, this capital plant continued to be an asset as the largest compressor plant within the Bristol Siddeley, Rolls Royce Company through successive take-overs. It was thus involved in the 'Olympus' development for the TSR.2 and 'Concorde', and in the 'Pegasus' development for 'Harrier' types, until the 1980s, when it was superseded by the major installation at Derby.

In 1954, two 'Comet 1' pioneer, jet-propelled airliners crashed catastrophically in the Mediterranean, with a cause eventually identified as metal fatigue of the passenger pressure cabin. The aircraft design was to recognised international standards and its failure was of world-wide consequence. Initial assumptions centred on wing-structure fatigue and the innovative propulsion with gas-turbine engines. The prototype VG had the equivalent of 6,700 flying hours and was under examination at RAE with cracks near the edges of the wing wheel-wells. Most recent, and complicating trials, had been the addition of 'Sprite' assisted rocket motors between the two 'Ghost' jet engines in each wing. Incidentally, I had been a passive observer on one of these trials, my first with jet propulsion and with the unusual additional bonus of rocketry. The proven fault, however, was definitely identified as that of the pressure cabin in aircraft flying at twice the previous operating height.

The engines were absolved, but trials in depth followed concerning a fresh feature consequent on the use of the gas turbine in flight: that of the gyroscopic couple of the fast-rotating rotor. The recovered wreckage of the 'Comet' G-ALYP which crashed near Elba included the major components of all four engines. Three had cracking fractures in their turbine mounting shafts, and the fourth had shed its bladed disk. It was concluded that the high couples were generated by the rotation in the fore and aft plane of the complete aircraft centre section after break-up. We designed and built a unique investigatory rig to the requirements of the stress office. A flight-standard Ghost-50 engine was mounted on a swinging framework as a pendulum arrangement and velocity meter, with its operational fittings and fuel supply. Before start it was swung upwards and restrained by a bomb slip. The engine was run at cruising rating, then released to swing in an arc under engine thrust and weight, and then restrained, with throttle shut, by a cable arrangement of two 'Vampire' oleo undercarriage legs. The bending moment cracking from the gyroscopic couple was duly reproduced on the unit under test.

Rocket Engines
Following my work in these developments I was appointed Test Plant Engineer – Rocket Engines, in support of the company's entry into manned aircraft, rocket propulsion work. Of particular interest was the fact that the oxidant selected was high strength hydrogen peroxide, described as 'HTP', at 85% concentration. Equipment and initial stocks were acquired from Germany under war reparations. Government initiative had set up the Rocket Propulsion Establishment at Westcott, Aylesbury, to the north east of Hatfield, where the Ministry establishment continued the development of the German wartime engines. De Havilland was to become the first large aircraft design and development company in British industry. In 1949, a DH-derived engine, the 'Sprite', using catalysed HTP to 600°C steam and oxygen, had been run at Hatfield on a simple emplacement test bed and then flown in trials on the 'Comet' prototype, G-ALVG, in 1951.

Dedicated rocket engine test houses were initially modelled on German designs. HTP manufacture was established by Laporte Chemicals at Warrington and nearby at Luton. Trials had been undertaken in 1957 at Hatfield with the fitment of German, former wartime, 'Walter' assisted-take-off rocket motors to a 'Lancastrian' aircraft undertaking trials of the DH 'Ghost' engines for the 'Comet'.

A 'Comet' airliner prototype takes off with 'Sprite' rocket motors, port and starboard.

My appointment coincided with the first running of the 'Super Sprite', a 'hot' derivation of the Sprite at 5,000 lbs thrust with the addition of fuel, which was being prepared for its first flight display at Farnborough in 1952. The next objective from this period was the new 'Spectre DSpe.1' engine with 8,000 lbs thrust for the 'SR.53' rocket-research fighter aeroplane, and its required engine test bed. The latter housed the engine in a relatively light but silenced structure with a blow-off roof and engine exhaust induced air-cooling streams. The manned control, instrumentation and operation room was housed alongside as a steel-reinforced, monolithic concrete chamber free to move under explosive impact in sliding, with a multi-layer, tarred paper surface, on an underlying separate concrete foundation. The design was a direct development from earlier German practice, the need having been demonstrated by a test cell accident at Westcott in which members of the German team transferring the technology had been killed. Rocket motor development tests continued to underline the hazards of early development, and its transition to manned use. The test bed itself was placed within a walled earth revetment and surrounded at a distance by dispersed HTP storage and other services. The HTP oxidant-related systems involved new technologies in materials and handling, demanding extreme cleanliness and purity. Next to the test bed we built the ground running arrangement for the SR.53 airframe rear-end, inclusive of the fin and high-mounted tail plane, allowing practical acoustic structural fatigue test as later engine development proceeded to the flight clearance standard.

Throughout this dynamic phase in activities, rocket work proceeded with the construction of other test beds and a wide range of rig houses and structures supporting other developments. These then assisted take-off variants for the 'Victor' and 'Vulcan' 'V' bombers, the later 'Spectre V' for the subsequent SR.177 for the RAF and German Navy, and the 'Twin Spectre' for the initial rounds in proving the 'Blue Steel' stand-off bomb for the V force. For this latter a peroxide drive auxiliary power plant was also developed by DH.

In the later 1950s, life in the aerospace industry seemed to exist on two planes; one of ongoing technical challenge, and the other in subjection to the pace of political change. The notorious Government White Paper of 1957 announced its intention to reorganise the military services and industry to accept that, in future defence, manned aircraft would give way to missiles.[1] As a consequence of this, various development and production contracts were cancelled. The DH Engine Company lost contracts for the Gyron Senior supersonic jet engine, the Spectre IV and V rocket engines, and other work associated with these developments. On top of this came the destruction of the second SR.53 rocket aircraft in a take-off accident at Boscombe in June 1958, with the death of Squadron Leader John Booth.

With the benefit of hindsight it may be observed that the Spectre series of rocket engines were the first to incorporate low-loss, internal turbines for propellant pump drive upstream of their combustion chambers, a practice now universal in high-performance engines. The Gyron Junior jet engine would continue as the PS.50 engine with the Bristol 188 steel supersonic research aircraft, and the PS.43 with 100 engines supplied for the 'Buccaneer' naval strike aircraft. The company's earlier jet engines continued in production, as did the Gipsy engines with the new Gipsy 215 supplying 100 for the 'Skeeter' observation helicopter.

Turbo-shafts for Helicopters

In 1958 I was appointed Engineer in Charge of the Overall Test Facility Department, with the company exploring new market opportunities. To replace the rocket engine activities, it capitalised on a long-term association with General Electric of America to licence development and production in Europe of the 1,000 hp T.58 helicopter turbo-shaft engine, in which GE had gained valuable contractual support from the US Navy. The company also found itself in a strong position in offering contract utilisation of the test facilities at Hatfield, with a particularly valuable liaison with Bristol Siddeley. This was mainly in connection with the 'Olympus' engine for the TSR.2 and for Concorde, and the 'Pegasus' for the Kestrel and Harrier.

Work with the American T.58 was originally to be complemented by a programme with their T.64 turboprop, but marketing stalled in the case of the latter. Initial runs of the T.58 were undertaken at Hatfield with the adaptation of a piston-engine matched Heenan & Froude dynamometer at 2,500 rpm to the 6,000 rpm of the turbo shaft. This arrangement utilised a Rolls Royce rig pairing of two 'Merlin' propeller gearboxes in a common casing running step up rather than down at twice speed: a nice example of industrial co-operation through the National Test Plant committee.

Most modern turbo-shaft engines are of the free turbine type where in effect a small jet-propulsion engine, running at high speeds, supplied its gas stream to a close-coupled, downstream turbine running at more conservative speeds and thence through gearboxes, classically with an output of about 6,000 rpm delivering to the aircraft transmission system and hence the lift rotor or rotors at, say, 300 rpm. The overall transmission system optimised for minimum torque weight, and with the combinations of multi-rotors and multi-engines, helicopter development is an art in the use of power plant and a prime field for electronic control.

In taking on the T.58, DH with the support of GE got on with development. The Westland 'Whirlwind' helicopter was first to be adapted to gas turbine power, with the conversion of the engine to the first use of electronic control. The engine itself was fully anglicised in the use of materials, and gearboxes were arranged with rolling bearings as against the GE preference for plain bearings. To speed the philosophy of transition to the gas turbine helicopter, DH became involved in gearbox design and development as a new field of endeavour, providing a new challenge to the test facilities.

Aerospace Rationalisation

The De Havilland Engine Company, as the last element of DH holdings to continue to trade as such, was taken over by Bristol Siddeley Engines in August 1962. I continued in post in my test work in closer intimacy with colleagues in Bristol. In June 1966 I was steered to pastures new, being appointed to the new post of Recruitment and Training Manager of our Divisional unit of 5,000 people at Leavesden, Stag Lane and Hatfield works.

Whilst my work was now less 'hands on', occasional forays into test work were still attractive. The DH company had earlier built the replica of the Wright 'Flyer' which is on display in the Science Museum in London, its engine – using an original casting – being built by the Technical School at Stag Lane. Its integrity test had been limited to running complete in operating back-driving the headgear of a lathe in the workshop. At this later period, the DH.51, precursor of the 'Moth' series, was being restored. Its eight-cylinder, air-cooled V 'Airdisco' 120 hp engine was rebuilt in the training workshop and required full Air Registration Board clearance. Dr Moult, retired Chief Engineer of Gipsy fame, had set it up for clearance on a test bed now operating at Hatfield. The bed lacked sufficient capacity in its dynamometer and surprisingly, without influence from me, one of my 'Stokes' dynamometers was pressed into service as a topping unit in clearance: this had been an earlier foray of mine into invention, being a water-pumping unit in a compact closed circuit inferring and indicating power by temperature rise in a constant flow secondary circuit. As a treat, at Leavesden aerodrome, I was later given the opportunity to hand-swing the engine of the aeroplane on starting its celebratory display.

★ ★ ★

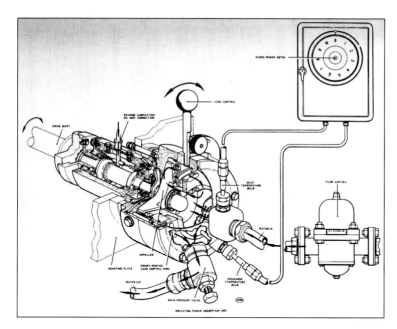

An explanatory diagram of a 'Stokes' dynamometer, evolved from a closed circuit pump-testing technique.

Family holidays in Cornwall had already aroused the mystique of the Cornish engine and its engine houses scattered across the landscape, so when I heard that the engines at Crofton on the Kennet & Avon Canal were being restored I took the opportunity of visiting the site on returning from a company meeting in Bristol. There I found a group of enthusiasts, largely from the Bristol Aeroplane and Engine companies, and was made welcome by them. My interest had already been stirred by reading L.T.C. Rolt on *George and Robert Stephenson* and *The Aeronauts*, and I find that my copy of *Inland Waterways* is touchingly dedicated, in my late wife's hand: 'All set for another fifteen years, March 14th, 1968'.[2] These were joined in 1972 and 1975 by the two masterpieces on Industrial Archaeology by Buchanan and Cossons respectively. So, having mustered in weekend attendance at Crofton and having served in the volunteer ranks in brick-cleaning and iron-scraping, I progressed to engine driving and then embarked on the commissioning phase of restoring the 1845 Harvey Cornish engine. This and the 1812 Boulton & Watt engine alongside it provide a sure foundation for appreciating the subtleties in the development of the Cornish engine, together with their pumping equipment, steam boilers, and ancillaries. The B&W engine, the oldest engine in the world still at work in its original house, represents the early practice in working at the relatively low lift of 40 ft on its power stroke. It was in later years converted to the Cornish cycle and pressure working. The 1845 Harvey engine, rather more demanding in its starting and operation, had originally worked as a Sims combined engine, whilst Cornish-built in a variation of the Cornish cycle. Both continue in operation to this day as an outstanding tourist attraction.

★ ★ ★

Meanwhile, in 1974 our Small Engine Division working on helicopter and general aviation propulsion had now been merged with the rest of the former Bristol Siddeley organisation within Rolls Royce plc. I was again job-rotated, in this case back from 'Staff' as Manpower Resources Manager, to the 'Line' post of Plant Engineering and Services Manager. The field was very interesting, and in so far as one of the immediate tasks required the replacement of some obsolete steam plant, my recent involvement with such plant was a recognised employment factor. The two main boiler houses were coal fired, and when the factories had been built for aircraft construction in the 1940s, they had been equipped with variegated 'Economic' two- and three-pass boilers from bomb-damaged factories, and represented the practice of the 1930s. The period of my tenure was immediately before that of the 'dash for gas' and the Ministry was then subsidising the purchase and installation of modern coal-fired boilers. That favoured was the Parkinson Cowan 'Vekos', a potentially dual-fuelled type projected as convertible from coal or oil to gas with alternative firing arrangements. Working and distribution pressures were at 80 psi, useful for our turbine spin rigs, then reducing for process and heating systems and back to the boiler houses.

The period of conversion took me up to the approach of my retirement, and during this period I was called upon to present papers to Mechanicals' Conferences, at the Science Museum and elsewhere; I was invited to give a series of lectures at Warwick University; and I enjoyed myself responding to requests for professional opinions, both at home and in the Middle East. In 1978 I visited the Helwan air base south of Cairo to give an opinion on plant engineering matters, which was of some delicacy because of the Arab-Israeli Wars. Interest on this trip was intensified on the journey out by travelling in a Swissair 'Trident' with its R.B.211 engines. Two take-off attempts were aborted at Zurich after trying to start a recalcitrant engine by wind milling. Such strange practice had been reported in the press, in the belief that with a three-engine aeroplane in winter conditions it could take off on two engines alone, a view that was definitely not recommended and was subsequently prohibited. The three-spool engine features no mechanical connection between its rotors, inhibiting the fan from energising the rest of the engine.

At the end of the Falklands War a group from Leavesden was invited to board the aircraft carrier HMS *Invincible* as she carried out deferred sea trials in Weymouth Bay. As ranking manager I led the group, and carried out the courtesies. The ship was at sea and we joined from a tender, individually hoisted aboard by a working party in exercise since the boarding gangway had been carried away in an accident with the same tender. This was a stout effort on the part of everybody, particularly for mature ladies of our party, but all quickly felt ship-shape, integrated and marinised after welcome coffee in the wardroom. The ship is, of course, Rolls Royce-powered, primarily with four 'Olympus' gas-turbine engines, the marine variant of the 'Concorde' engine, in this case geared in pairs to her twin screws. The prime purpose of testing was noise signature, so she steamed about the bay energetically in a fair sea state. Whilst her size dampened excessive motion, her turning rate caused distinct banking. Ours was a mixed group of technical and administrative staff, and we appeared to be given a complete run of the ship, from the bridge to machinery spaces. A grand day out!

Antiquity beckons

My interest in Industrial Archaeology, engendered by Tom Rolt's evangelising and the experience of Crofton, continued to grow and I became actively involved in the project to restore the Kew Bridge engines in London. The surviving Cornish cycle engines at Kew were widely admired, the site having been designated in 1944 as a museum by the Metropolitan Water Board, which had preserved the engines although boiler capacity had been removed as there was no intention for further operation. With the new enthusiasm of the 1970s, however, Kew became a prime objective for the restoration of steam working. With other members of the Crofton team, I was drawn to the project and the new group of supporters undertook a large but eventually successful restoration, to become a most attractive working museum.

The Chief Engineer at Crofton, Ron Plaster, and the General Manager, Tony Cundick, both transferred their skills to Kew, and Nick Reynolds accepted the role of Secretary. Work began on the earliest engine, the 1820 Boulton & Watt 64-inch cylinder beam engine. The engine had been built in Birmingham for the Regent's Canal Company in pumping water from Chelsea on the Thames adjacent to the Military Royal Hospital, to the confluence of the new Regent's Canal with the Grand Junction Canal at Paddington. The engine is redolent of the mystique of the early steam engine, having been up-dated to the new Cornish technology of high pressure working in its early years. Its working requirements also changed from canal to domestic water supply pumping.

Over the years, other engines have been brought to restoration at Kew Bridge, the most notable being the 90-inch cylinder Copperhouse Cornish engine, built for operation in 1846 to the design of Thomas Wicksteed, in the works of Sandys, Carne and Vivian in the area of Copperhouse, at Hayle in Cornwall. This large engine, as it is worked for demonstration at Kew, symbolises both the achievement of Cornish mining techniques and of London's water supply.

Taylor's 85-inch engine built by Perran Foundry to the design of Michael Loam for installation at United Mines, Gwenap, Cornwall. (drawing by Max Millar, reproduced by courtesy of the Institution of Mechanical Engineers)

The overall collection now represents various phases of steam engine history, with the late nineteenth-century 'Pulsometer' pump replicating the operating cycle of the 1698 'Savery' pressure/vacuum engine, as developed from that of the Marquis of Worcester which is regarded as the first steam prime mover. The 'Maudslay' beam engine, commissioned for the site in 1838 as the firm's first engine following the Cornish cycle, was restored in 1985. The last restoration at Kew has been that of the unique 70-inch Harvey/Bull engine, but with the even larger 100-inch Harvey engine, a structural crack in the beam has led to a decision to preserve it but not to try to restore it fully.

Conclusion

I regard myself as fortunate in having enjoyed a working life formed primarily in the field of aero-engine testing, but being able to complement this by practical involvement in industrial archaeology through participation in the restoration of steam engines at Crofton and Kew. Amongst many highlights, I recall taking the opportunity on one weekend Sunday steaming, of having started and driven both engines at Crofton, and then hastening to London to drive both the Boulton & Watt and the Copperhouse 90-inch engines in the afternoon: a reward for volunteer service at both sites.

With restoration completed to standards that allow both understanding and operation of a range of historic engines, it is interesting to find this willingness to learn from the technological achievements of our predecessors extrapolated into other areas. In the aero-engine field, for example, there is the remarkable saga of the 'Moth Club' in continuing to fly DH Moth aeroplanes with their Gipsy engines, thus carrying on experience from the 1920s into the twenty-first century. Similar examples of restoration and operation by enthusiasts can be found in motorcars, steam traction engines, rail locomotives, and steam ships. As exemplified in British experience, this apparent reversion to the days of the former industrial economy demonstrates a willingness to capitalise on a receding skill base.

Such has certainly been the case with those of us who have shared in the experiences of restoration and steam working at Crofton and Kew. With myself and my peers now mature observers, it has been possible to undertake retrospective historical tests with the support of local universities, buttressing the study of their students and staff in technological history. At Crofton, two sets of engine trials in the classic mode of the thermodynamic laboratory have been undertaken: the first, in 1949, predated the restoration, and the second, in 1998, made it possible to compare results with performance in the restored condition. Both trials were undertaken under the auspices of the Newcomen Society: that of 1949 under the guidance of Dr N.G. Calvert of Liverpool University, and that of 1998 with his son Dr J.R. Calvert from Southampton representing the approach of today.[3]

At Kew, test investigation has been directed at the practical investigation of the problem presented in 1862 by the cracking of one of the dual beams of the 100-inch Harvey engine next to the 90-inch engine. John Porter and John Loadman conducted this study and reported their findings in a paper to the Newcomen Society in 2007: 'The

Cast Iron Beams at Kew Bridge Pumping Station – Faithful Friends or Fickle Enemies?'[4] As well as 'indicating' the engine steam cycle, the beam was strain-gauged against working criteria and the tests were analysed inclusive of a finite element structural study. Thus twenty-first-century computer aided engineering was brought to bear on consideration of nineteenth-century problems in the strength of materials.

I hope these thoughts on the subject of experience of testing in modern aerospace and in obsolete steam prime movers, will convey something of the value of historical studies to the understanding of modern technology. They represent, at least, the pleasure and significance of such relationships in the experience of one enthusiast for *Landscape with Technology*.

Notes
1. The Defence White Paper (Cmd.230), 1957: proposed by Duncan Sandys, this passed into government policy as a considerable setback to the defence industry.
2. L.T.C. Rolt, *The Inland Waterways of England* (London: Allen & Unwin, 1950).
3. *Transactions of the Newcomen Society*, 71, 1999-2000.
4. *Transactions of the Newcomen Society*, 77/1, 2007: John Porter and John Loadman, 'The Cast Iron Beams at Kew Bridge Pumping Station – Faithful Friends or Fickle Enemies?'

8: The New Great Space Race

David Ashford

In 1961, I was hired straight from university to work with the Hawker Siddeley Aviation Advanced Projects Group at Kingston. My first job was in the hypersonics team. Among the projects we investigated were spaceplanes – aeroplane-like vehicles capable of flying to and from orbit to launch a satellite or supply a space station. The big idea was to replace the converted ballistic missiles then in use with vehicles that could fly more than once. In principle, this involved adding wings, tail, cockpit, landing gear and heat shield to an otherwise expendable vehicle. One does not need to be a rocket scientist to appreciate that an aeroplane is a much more practical and economical form of transport than a ballistic missile.

Most large European and US aircraft companies had similar ideas. By the mid-1960s there was a consensus that spaceplanes were the obvious next step in space transportation and were just about feasible with the technology of the time. The X-15 rocket-powered research aeroplane was making regular flights to space, albeit on a sub-orbital trajectory that gave but a few minutes above the atmosphere. Several papers at the time predicted that spaceplanes would lead to an airliner service to and from space, reducing the cost of orbital transport by several orders of magnitude.[1] This would lead to a new space age, although that expression was not used at the time.

Well, it has never happened. To this day, all spacecraft have been launched by vehicles with large complex throwaway components. How has this come about? To answer this question, we need to look at history. The first satellites were launched using converted ballistic missiles rather than rocket-powered aeroplanes because the latter would have taken longer and cost more to develop. Due to the pressures of the Cold War, the first men in space also got there on top of ballistic missiles, and the use of expendable launchers persisted during the 1960s race to the moon.

The next major project was the Space Shuttle, the design of which started in the 1970s. The advantages of spaceplanes were by then widely appreciated and the early designs of the Space Shuttle were indeed fully reusable. Budget pressures then forced NASA to choose between a smaller reusable design, which would have introduced the aviation approach, or giving up on full reusability. The habit of expendability was by then strong enough for NASA to choose the latter. The largely expendable Space Shuttle is as expensive and as risky as the vehicles that it replaced. It has a cost per flight of about one billion dollars, so very few flights can be afforded.

This history has created institutions with ways of thinking which have repeatedly reinforced the throwaway launcher habit. This mindset is the largest obstacle in the way of spaceplane development. Even today, NASA, ESA and other space agencies are developing new expendable launchers.

It is worth considering in more detail precisely why the Shuttle is so expensive to fly. After all, the Shuttle Orbiter – the component that carries the crew and that lands back on earth – is fully reusable. The basic cause is that the complete Space Shuttle vehicle is not reusable – the Orbiter is launched using an expendable propellant tank and recyclable (not reusable) solid rocket boosters. This leads to three main reasons for the high cost. First, the total cost per flight of the complete Shuttle is inevitably so high that the number of flights remains too low for the Orbiter even to approach airliner standards of maturity. Even with mass production, the cost per flight of the Shuttle would be far too high for potential new commercial space markets. A new airliner typically requires 1,000 test flights over a one- to two-year period before it is allowed to carry passengers. The Shuttle has made some 120 flights in 27 years.

Second, the safety of the Shuttle crew depends on large, complex, non-reusable components, which are inherently unsafe by conventional aviation standards. Indeed, the two fatal accidents were caused by the non-reusable components failing in a manner that damaged the Orbiter. It is not much of an exaggeration to say that every time the Shuttle flies, a crew of seven is flight-testing a ballistic missile.

The third reason is that each flight has a different payload that requires close integration with the Orbiter.

This technical immaturity, concern for safety, and use of non-standardised payloads, results in the number of people involved in flight preparation being measured in thousands, compared with the ten or so for an airliner. Spaceplanes, however, are fully reusable and therefore not subject to these limitations.

However, all is not lost. The private sector has taken the lead. The first privately funded spaceplane, SpaceShipOne (SS1), reached space in 2004. Virgin Galactic plan to start carrying passengers on brief space experience flights in a few years time in an enlarged development of SS1. Several other companies, including my own, are developing spaceplanes. The main obstacle is obtaining the funding, largely because spaceplanes are not yet in line with space agency policies.

Present designs are sub-orbital in that they can fly fast enough to zoom up to space height for a few minutes but not fast enough to stay in orbit. Orbital spaceplanes need some six times the speed of sub-orbital ones and will cost some ten times more to develop. This cost is at present beyond the means of the private sector. The best way ahead is probably a public-private partnership. Government would pay for developing prototypes adequate for their own purposes. The private sector would then take over commercial exploitation. Governments would save money on present space programmes alone. The cost of the first lunar base, for example, would be about ten times less than with present plans that use large new expendable launchers.[2] Space agency avoidance of serious discussion of this alternative is little short of scandalous.

Perhaps surprisingly, the UK is best placed to break the mould of thinking on space transportation. We have all the technology for an entry-level spaceplane; we have a world-class aircraft industry; and we are the only major industrial country without significant interests vested in 'old space'. If we start down the winning strategy, we should be able to persuade international partners to join in a programme of further development. In this way, the UK could become the centre for a large new European spaceplane industry, with considerable political, economic, and educational gains.

The rest of this paper considers two key aspects of this situation in more detail. First, I will consider the potential of spaceplanes to reduce launch cost. Then I will show how the UK could take the lead. There will be some repetition from the above summary.

Rockets, Aeroplanes, and Spaceplanes

It is relevant to consider in more detail why present-day spaceflight is so expensive. Fig. 1 compares expendable launchers with airliners. The cost per seat to orbit in a launcher is about four orders of magnitude higher than a seat in a long-distance airliner flight. The main technical difference is that an airliner can be flown tens of thousands of times whereas a launcher can be flown once only. The other technical differences (vertical take-off, rocket engines, lack of pilots, and number of stages) have some effect on cost but are not nearly as significant.

The main operational difference is that airliners make some ten million flights per year in total whereas launchers make fewer than 100. This follows from their high cost, which in turn follows from expendability. Expendable launchers are so expensive that they are suitable only for government missions and for the few commercial applications that can afford such high costs, especially satellites for communication. There are about 20 launcher types on the market, so the average number of flights per year for each type is less than five. Not very efficient!

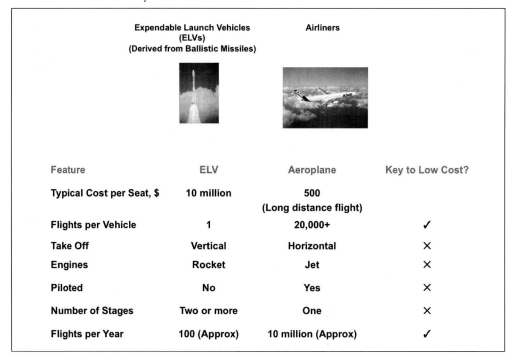

Fig. 1. Launchers compared with airliners.

Present-day airliners of course cannot fly to and from orbit. So how will spaceplanes compare in cost with airliners? Figure 2 shows some estimates for a spaceplane with a take-off weight comparable to that of a large airliner. We have assumed a technology level such that prototypes could be built in five to seven years and that the design has then matured following a long production run with continuous product improvement.

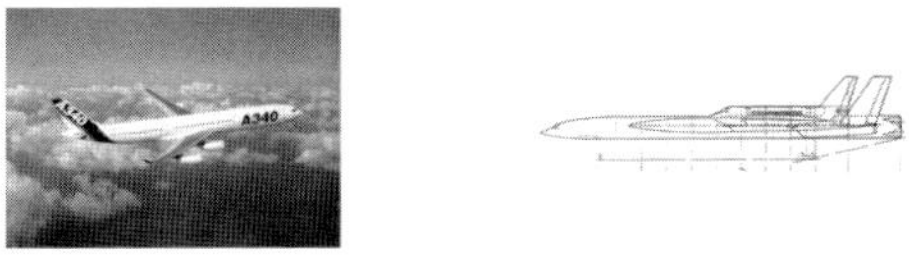

	Airliners	Spaceplanes (When Fully Developed)
Flights per Vehicle	20,000+	20,000+
Cost per Flight, $million	0.2	0.5
Number of Seats	400	50
Cost per Seat, $ (Typical)	500	10,000

Fig. 2. Spaceplanes compared with airliners.

When fully developed, there is no obvious reason for spaceplanes to have a shorter life than airliners. They will cost several times more per flight because of greater complexity and because they will probably use hydrogen fuel, which is expensive. They will carry fewer seats because they have to carry more fuel. The resulting cost per seat will be, very roughly, 20 times mores than a seat in a long-distance airliner flight, at around US$10,000. This is some 1,000 times lower than the cost today.

When these low costs are achieved, public access to orbit for business or leisure will become routine and affordable. There are millions of people in the industrialised world who would be prepared to save for the trip of a lifetime to a space hotel for a few days. In effect, we will have a new space age.

Science is a major application of space. Our knowledge of the universe has been transformed by information from satellites clear of our distorting atmosphere, by space probes, and by manned visits to the moon. There is a useful analogy with science in Antarctica, where much useful geophysical science has been carried out. Antarctica is isolated and it costs of the order of £10,000 to send a scientist there. If the only available form of transport was ballistic missiles, it would cost of the order of £10,000,000 to send someone there, and you can imagine how much less science would have been carried out. Well, that is how it is with space science. As soon as spaceplanes replace expendable launchers, we will see a new golden age of astronomy.

It must be emphasised that costs that low depend on fully mature spaceplane designs. Early prototypes will cost far more. These could be built in five to seven years, and it would then take about eight years to mature the designs, especially to develop a long-life rocket engine. Thus, a 1,000 times cost reduction appears to be possible within 15 years. Achieving this timescale depends on large new markets growing rapidly to provide economies of scale, especially public access to space.

How the UK can lead

The main obstacle to achieving the new space age soon is mindset. Even now, NASA is planning to build large, throwaway launchers for a new programme of lunar exploration, even though, as mentioned earlier, it can readily be shown that costs could be reduced about ten times by adopting an aviation approach. The UK is well placed to be the first to break the mould of thinking because we are the only major industrial country without major commitments to 'old space' projects. We also have a world-class aircraft industry with access to all the technology required for an entry-level spaceplane. Of all aeroplanes that have actually flown, the best technology demonstrator for an entry-level, sub-orbital spaceplane is arguably the Saunders Roe SR.53 rocket fighter, which first flew in 1957!

Fig.3. Saunders Roe SR.53 rocket fighter, 1957.

When it was cancelled in 1958, Saunders Roe proposed a space research variant. The Ministry showed some interest, but not enough to make it happen. What might have been! My own company's entry-level spaceplane project, Ascender, as shown in Figure 4, is in effect a simplified and updated SR.53.

Ascender would be useful for carrying science experiments, high-level photography, meteorological research, astronaut training, and carrying passengers on brief space experience flights. Perhaps more importantly, it would pave the way for the first orbital spaceplane. Our Spacecab (Figure 5) has been designed specifically to be the most competitive candidate for the first orbital spaceplane. It is in effect an updated version of the 1960s European Aerospace Transporter project designed to minimise development cost by using existing technology. The difficult part of Spacecab design was avoiding anything difficult!

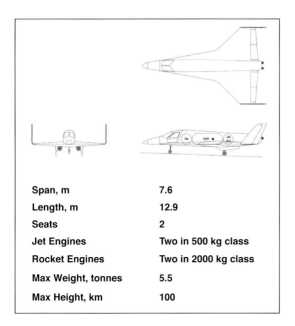

(left) Fig. 4. The Bristol Spaceplanes Ascender entry-level spaceplane.

Span, m	7.6
Length, m	12.9
Seats	2
Jet Engines	Two in 500 kg class
Rocket Engines	Two in 2000 kg class
Max Weight, tonnes	5.5
Max Height, km	100

(below) Fig. 5. The Bristol Spaceplanes Spacecab.

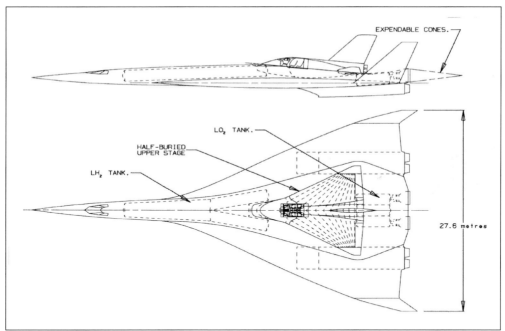

Spacecab has a payload in the one tonne class. This could be a satellite or supplies or crew for a space station. Spacecab has two stages so that existing technology can be used. (A single-stage vehicle using the most efficient rocket engines available would require 87% of take-off weight to be propellant, which is impracticable for a robust design. This can be reduced by using hypersonic air-breathing engines, but these need an extensive research and development programme.)

Both stages are piloted and take off and land horizontally so that it is as much like an airliner as practicable. The Carrier aeroplane is about the same size as Concorde. However, it is far simpler because it has to maintain its maximum speed for long enough only for the Orbiter stage to separate (a minute or two) rather than to cross the Atlantic (more than two hours). The Carrier aeroplane uses existing jet engines for acceleration to Mach 2 and then rocket engines up to Mach 4. During the rocket part of the acceleration, Spacecab climbs to near-space so that stage separation can take place where the air and thermal loads are manageable.

Spacecab could be developed at surprisingly low cost because:
- Prototypes can be developed in experimental workshops. (This is not possible for man-rated, expendable launchers because of their inherently poor safety.)
- These prototypes can be used for pioneering operations before they have been fully certificated to airliner safety standards (which would increase development costs by about ten times) because they would be far safer than the expendable launchers that they would replace.
- For early designs, ahead of most of the competition, great efficiency is not required and so relatively generous design margins can be used. The early designs can then be improved as the market requires.

As soon as the first orbital spaceplane enters service, it will be able to undercut any expendable launcher of comparable payload. This will encourage higher traffic levels, which will in turn release investment to mature the design. This will lead to even lower costs and higher traffic levels, and so on down a virtuous cost spiral until the lower limit of spaceplanes using mature developments of existing technology is reached. As we have seen, this is about 1,000 times lower than the cost today and could be approached in about 15 years. In this way, it is all but inevitable that the first orbital spaceplane will lead to an airliner service to and from orbit and to a new space age.

Such low cost to orbit will enable the cost of other space vehicles to be greatly reduced, such as space stations, lunar transfer vehicles, lunar landers, lunar bases, and space probes to more distant parts.

So, we are talking revolution rather than evolution, with the UK well placed to play a leading part.

Conclusions

A new space age is in sight, based on aviation standards. This will have greatly reduced costs and improved safety. The main obstacle is the scandalous failure over the past few decades of government space agencies to take this prospect seriously. However, courtesy of private sector initiatives, progress at last appears to be possible. The winner of the new great space race will be the first major player to back the aviation approach to space transportation.

Notes

1 Cornog, R., 'Economics of Rocket-Propelled Aeroplanes', *Aeronautical Engineering Review*, September 1956; Dornberger, W.R., 'The Rocket Propelled Commercial Airliner', University of Minnesota, Institute of Technology, Research Report No.135, November 1956; Koelle, H.H., 'Assessing Re-usable Space Vehicles', *Astronautics and Aeronautics*, June 1964; Ashford, D.M., 'Boost Glide Vehicles for Long Range Transport', *Journal of the Royal Aeronautical Society*, July 1965.
2 Ashford, David, 'The Aviation Approach to Space Transportation', *Journal of the Royal Aeronautical Society*, August 2009; Ashford, David, *Spaceflight Revolution* (London: Imperial College Press, 2002).

9: Working Historic Machinery – can it be safe? The Case of Crofton Pumping Station

Geoff Wallis

Machines are devices designed to perform a useful task. When the task changes, the machine wears out, or when advances in technology make it obsolete, the machine becomes redundant. No longer earning its keep, the machine's value may be reduced to that of its materials, house, or the land it occupies, together with any perceived cultural value. The need to reuse the space it occupies or to realise its monetary value means that most redundant machines are scrapped. The minority that manage to survive may eventually experience a second lease of life as 'heritage', which may present the opportunity for it to be restored to a working condition, but inevitably increases the risk of injury to those who attend or visit the machinery, the numbers of whom may be well in excess of those it encountered in service.

Can an historic machine, built during the years before the culture of 'health and safety' developed, be operated safely nowadays? If so, can safe operation be achieved without compromising the preservation of the heritage values of the machine? This study assesses some practices and why they have taken place, with reference to a particular site.

Crofton Pumping Station near Marlborough, Wiltshire is the world's oldest working steam engine still on its original site performing its original function. The Station's historic equipment is relatively complete, set in a landscape relatively unaltered in 200 years, and bears eloquent witness to the dramatic changes in transport and technology that took place in the nineteenth century. It is a Grade 1 Listed site of world importance.

The Pumping Station has been under the control of the Kennet & Avon Canal Trust for 40 years, cared for by a succession of volunteers, each generation of which has wrought changes. In the absence of clear, written policy guidance, volunteers have done their enthusiastic best, but conservation has been *ad hoc*, and sometimes below the standards achieved on sites of comparable historic importance. This study explores an alternative approach for Crofton, and recommends that changes be implemented before 2012, the bicentenary of the commissioning of the first Crofton engine.

Industrial Preservation – a brief historical perspective

The modern industrial preservation movement has its roots in the post-Second World War period when sites damaged during the war, or made redundant by advancing technology, were cleared rapidly as 'modern' Britain was born. By the 1960s this process was well-advanced and the realisation developed that, whilst forward-looking change was good and necessary, a balance was also needed to retain worthy examples of the past technology on which the Empire had been built, and which had helped Britain survive two world wars.

The foremost champion for the preservation of historic machinery is Sir Neil Cossons, OBE, MA, FSA, former Chairman of English Heritage where for seven years he was responsible for the conservation of all England's listed buildings. He is a former Director of the Ironbridge Gorge Museum, The National Maritime Museum and the Science Museum, a specialist in industrial archaeology. He asserted recently:

> It is one of the happy paradoxes of the last 30 years that much of this country's history that otherwise would have been lost has been secured by a new type of museum working outside the established framework of public institutions. The vision and energy, focus and sound management of these independent museums represents an extraordinary national achievement, of people with fire in their bellies and unwilling to take no for an answer, whose work future generations will come increasingly to value and respect.[1]

The impetus for the preservation movement came from amateurs. They were railway buffs interested in preserving railways and rolling stock; aircraft workers saddened to see pioneering aeronautical hardware scrapped; mariners; wind and watermill enthusiasts; and engineers with a passion to preserve historically important engineering installations such as Crofton Pumping Station, the focus of this paper.

After the initial 'reclamation' stage was over and machinery was working, the challenge was to operate and maintain in a sustainable fashion what had been restored. Different skills were needed, including publicity, marketing and interpretation. Often the original restorers moved on and were replaced by others who perhaps had little knowledge of the site's earlier appearance and ambience. They wrought changes, and were themselves eventually replaced.

Thus on some sites layer upon layer of changes took place, but increasingly thought was being given to this process and, more generally, to the philosophy behind conservation. This was partly as a reaction against indiscriminate change, and partly from a desire to provide a sound philosophical basis for a worthwhile activity.

Long before this period conservation policies and practices for the treatment of classical archaeological finds, books, paintings, clocks, scientific instruments, and museological artefacts had become well-developed. Then in 1979 the Burra Charter was agreed by international treaty, and gradually conservation policies were developed by the National Trust, English Heritage, CADW, Historic Scotland and other organisations responsible for Britain's built heritage.

Codes of practice developed, written and unwritten, setting out how best to treat historic architecture, but working boilers and steam engines presented new uncertainties. *How* can you preserve an historic boiler when its plates have been condemned as unserviceable by an all-powerful boiler inspector? And *why* should one consider doing so? Preservation often meant extensive replacement and non-reversible intervention. How far should one go? Could we afford *not* to restore to working order so as to generate interest? Could boilers and steam engines survive into the future, static and unused, in buildings that are expensive to maintain?

Official bodies such as the national museums, governmental heritage agencies and the National Trust adopted a preservationist approach. Most decided that historic machinery would ideally be kept in working condition provided it would not thereby be spoiled or put at unacceptable risk. Machinery would be preserved with its building and surrounding site to retain the appropriate context. Intervention would be kept to an absolute minimum and be reversible wherever possible. Visitor access would be encouraged but in a way that impacted minimally on the historic machinery itself. Traditional materials, technology and practice would be used where visible, and unavoidable modern material would be kept hidden as far as possible.

Increasingly these standards have become enshrined in planning law, clarified by Planning Policy Guidelines. These tenets have become widely accepted as the 'industry standards' to which responsible owners and operators of historic sites aspire. The National Trust has been one of the pioneers in defining and applying good conservation practice, and now has strict in-house procedures for ensuring that best practice is applied to their buildings, machinery and portable artefacts.

In 1968, at Crofton Pumping Station, four engineers from the aircraft industry – Roy Simmons, Ian Broom, Ron Plaster and myself – recognised the need to preserve this unique site. They started the process of preservation with their own limited resources and gradually drew in other supporters. With 'fire in their bellies' they did what they could, guided by a senior engineer from the water industry and Newcomen Society member, Arthur Pyne. They were motivated by a desire to see a threatened installation continue its existence with minimal change. However, their aims have only partially been achieved.

Historic Buildings Preservation – the legislative context

Under British law, everybody at each echelon in any organisation owning or managing an undertaking has a duty of care for the safety of everyone on the site irrespective of whether they are employees or volunteers.

The operation is subject to a range of laws including health & safety legislation enforced by the Health and Safety Executive, environmental health legislation enforced by the local authority, and fire regulations enacted by the local fire authority.

Additionally on historic sites, conservation legislation applies, enforced by the local authority with advice from English Heritage and statutory consultees, such as the Georgian or Victorian Societies.

England's 500,000 most important historic buildings are listed as being 'of architectural or historic interest'. The 2% of listed buildings which are of special importance are designated Grade 1. The primary legislation governing the preservation and development of our built environment is *Planning Policy Statement 5: Planning for the historic environment*, commonly referred to as PPS5. This is the 'rulebook' by which local authorities and their consultees work.[2]

Crofton Pumping Station is a typical site with historic working machinery open to the public. It is an undertaking where paid and unpaid workers carry out tasks at the direction and supervision of the local branch management, acting with devolved responsibility and authority from the Kennet & Avon Canal Trust Council, and ultimately from the Trustees. So, as a place of work Crofton is subject to the *Health and Safety at Work Act* (1974). The site also sells food prepared and consumed on the premises so it is subject to *The Food Safety Act* (1990), and like all sites it is subject to fire regulations.

This legislation is primarily concerned with the health, safety and well-being of human beings. But as a Grade 1 listed site, Crofton Pumping Station is also subject to the conservation legislation enshrined in the *Town and Country Planning Acts*, primarily concerned with the historical integrity of the site, its context and contents.

These two very diverse focuses in legislation mean that conflict is inevitable in deciding on a legal course of action, particularly as there is no clear priority in law for conservation over human welfare, or vice versa. This conflict places all involved in this historic site in a dilemma from which there appears to be no escape. Failure to comply with all applicable legislation is a criminal offence, punishable with a fine or imprisonment, actions being brought by an enforcing agency acting in the public interest. Additionally, in the case of personal loss or injury, litigation may be pursued for damages, awarded by the courts according to the severity of the injury and the degree of culpability of each party.

Individuals making, enforcing, or executing decisions in the Kennet & Avon Canal Trust and the Crofton Branch can therefore be held personally responsible for failures if found negligent, and their organisation can also be held responsible corporately. In its outworking, health and safety legislation is inevitably often in conflict with conservation legislation, but the conflict is an unequal one. Firstly, there is far more legislation relating to health and safety than there is to conservation, demanding and receiving greater attention. Secondly, the threat of prosecution for failure to comply with safety legislation is perceived to be greater than the threat of non-compliance with conservation legislation, and the penalties more severe. A cash-strapped local authority is considered unlikely to take costly action to enforce conservation legislation, whereas the Health and Safety Executive would do so to enforce safety regulations. Thirdly, the risk of litigation for personal injury is real, and damages potentially large, whereas the risk of litigation by an offended amenity group is small, and uncertain in its outcome. Injured people sue, injured buildings don't.

In the management of historic buildings it is regrettable that the predominance of safety over conservation often results in an imbalance detrimental to the building or its historic contents, a situation not required by law but driven by the perceptions and fears of those subject to it. It is a duty of bodies responsible for the care of important historic structures to recognise this imbalance and take effective steps to ensure that the interests of sensitive sites are fully represented. Local authorities have a duty to ensure that this balance is achieved and can influence action by providing advice, in consultation with English Heritage specialists as necessary, by ensuring grants are given only where proper conservation plans are in place and actioned, and by enforcing conservation legislation or at least threatening to do so.

Conservation principles in practice

English Heritage's recent draft *Conservation Principles, Policies, and Guidance* proposes that the significance of an historic site may be articulated as the sum of its heritage values considered under four headings:

Evidential values: the potential of the fabric of the building to yield primary evidence.
Historical values: the way in which the building provides a means of connecting the present to past people, events, and aspects of life, both by illustrating important aspects of canal and social history, and through its association with notable people and events.
Aesthetic values: the way in which people derive sensory and intellectual stimulation from the building
Communal values: the meaning of the building for the people who identify with it and whose collective memory it holds.

The Paul Drury Partnership's *Rochester Castle, Conservation Plan* (Medway Council, October 2009) adopted these definitions and graded the many attributes of the Castle to produce a rigorous and detailed assessment of its historic significance, using the following gradings:[3]

A Exceptional significance: elements whose values are both unique to the building and are relevant to our perception and understanding of it in a national and international context. These are the qualities that warrant listing as grade I.
B Considerable significance: elements whose values contribute to the building's status as a nationally important place. These are the qualities that justify statutory protection at national level.
C Some significance: elements whose values make a positive contribution to the way the building is understood and perceived, primarily in a local context.
D Little significance: elements whose values contribute to the way the building is perceived in a very limited but positive way.
N Neutral significance: elements which neither add to nor detract from the significance of the building.
INT Intrusive: elements of no historic interest or aesthetic or architectural merit, that detract from the appearance of the building, or mask our understanding of it.

Paul Drury's work offers a good template for assessing the significance of any historic site, and a detailed 'significance statement' should form the basis of any strategic planning on conservation of historic working machines.

Dorothea Restorations Ltd Conservation Policy Statement

One of the longest-established firms of practical conservators working on historic machinery is my own company, Dorothea Restorations Ltd. Established in 1975 it soon became clear that clients generally sought the company's advice as to 'what should be done' rather than specifying the work themselves which they wanted to be carried out. This promoted much discussion amongst the directors, who agreed that the company's conservation philosophy should be developed and written down. By 1988 this had reached its third draft, comprising four sections.

The first set out definitions, experience having showed that discussion degraded from matters of conservation into semantics until agreement could be reached on the meaning of the terms used. The second was an early attempt to define historic importance, particularly as applied to historic machinery. Then the company's fundamental aims were enunciated, and finally a few good conservation practices were defined. A summary of the statement follows:

Definitions:
Artefact means machine, structure, instrument or other piece of man-made equipment.
Historic importance means aesthetic, scientific, technological or social value for past, present or future generations.
Material is the physical substance of which an artefact is made.
Preservation means maintaining the materials of an artefact in their existing state and halting deterioration.
Restoration or Repair means returning the existing materials of an artefact to a known earlier state with minimal introduction of new material.
Reconstruction means returning the existing materials of an artefact to a known earlier state involving more than the minimum amount of new material.
Maintenance means the continuous care of an artefact without alteration of its materials.
Adaptation means modification of an artefact to fulfil a new compatible use.
Compatible use is one which involves no change to the historically important features of an artefact. In less than ideal circumstances a compatible use may be one which involves reversible and minimal changes.
Conservation refers to all the processes of preserving artefacts of historic importance, such as preservation restoration, repair, maintenance, and where of an acceptable standard, reconstruction and adaptation.

Historic Importance must be precisely defined as the basis for selecting conservation processes. An artefact's aesthetic, scientific, technological and social merit is assessed in terms of:

Age.
Uniqueness of design, scale, materials, etc. when originally constructed.
Rarity as a survivor of its type.
Evidence of past style, design, innovation, usage materials, constructional practice, etc.
Association with persons, places or events.
Spiritual, political, or cultural significance for a particular group in society.
Exceptional aesthetic qualities of form, colour, decoration, etc. and part played in the immediate environment or landscape.
Condition and extent of remaining original material.
Service life - does it still perform its original operation, or can it be made to do so?

The Fundamental Aim is to *Preserve*, to halt the processes of deterioration and stabilise condition. This recognises the intrinsic worth of the artefact's materials, and ideally should be the limit of conservation work.

Where stabilising its condition is insufficient on its own to ensure an artefact's long-term survival, conservation must be selected to ensure minimal disturbance to the artefact, and that reversible processes are used whenever possible.

These simple, fundamental principles recognise the need to preserve the artefact or machine intact as evidence for future generations.

Good practice
Good practice derives from the principles set out above, and will include improvement of the immediate environment (protection from the elements, covering, packing, etc.), dehumidification, reduction of

pollutants, alteration/down-grading of service requirements, use of replicas for working equipment, relocation to a less destructive environment if this is proven to be the only means of ensuring preservation. It may involve consolidation of existing materials (only where unavoidable to ensure survival, e.g., paint, friable timberwork), and use of additional materials or structure, not fixed to the artefact, for additional strength or support. Where original coatings are ineffective or cannot be reclaimed, new protection may be required in the form of removable coatings such as wax, or permanent ones like paint, inhibitors, etc.

For repairs, traditional materials and techniques should be used, provided these can be distinguished on close inspection from originals on completion, and for years to come. Where parts or materials are removed, whatever their condition they should be tagged and stored adjacent to the artefact from which they came.

Ongoing maintenance requires periodic planned inspection by competent, diligent persons, and the provision of comprehensive written maintenance schedules. Where machinery is operated, guidelines should be written and staff trained in routine operation and emergencies.

All conservation work should be properly recorded before, during and after, measured drawings and photographs being particularly useful. Records are, however, useless unless held in secure and stable long-term storage, indexed so as to be accessible.

Forty years of change at Crofton Pumping Station

Crofton Pumping Station has been in the ownership of the Kennet & Avon Canal Trust for over 40 years, during which time the engines have been preserved and still work regularly, the houses have been maintained, the site is reasonably well known and attended, the operation is financially viable, there is considerable volunteer interest, and the site is run efficiently and safely.

The army of volunteers who have kept the site in operation over the years and turned it into a major tourist attraction are to be recognised, thanked and congratulated. But many changes have taken place that were not minimal, unavoidable, or reversible, so that by modern conservation standards the toll on the building and environs has been too great. Historic authenticity has been compromised, often unnecessarily.

Whilst it is accepted that some change to historic sites is necessary, modern conservation practice aims at minimal intrusion, and therefore subscribes to a presumption against changes where avoidable. Much of the site's original simplicity and austerity have been lost, giving a false impression of what a nineteenth-century, rural pumping station was like, disappointing visitors expecting to experience a traditional ambience. Insensitive or ill-informed treatment suggests poor stewardship of an important Grade 1 listed site, which can reflect badly on both the Branch and Trust.

Sir Neil Cossons is a specialist in industrial archaeology who has known Crofton Pumping Station well for over 40 years. He observed:

> Like a number of similar places, the pumping station at Crofton, built in the early years of the 19C to lift water into the K & A Canal, and a site of outstanding importance, has succumbed to changes driven not by the needs of conservation so much as public circulation, health and safety regulation and through the introduction of items of alien and unrelated equipment justified on the spurious grounds that greater public interest might be generated and the volunteers will need something additional to do.
>
> The absence of conservation plans, informed and useful advice, or qualitative conditions applied to such public funds as are available makes these sites especially vulnerable to levels of compromise that would be unacceptable in other fields.[4]

Crofton Pumping Station

Nick Reynolds, past Treasurer to the Kennet and Avon Trust and prime-mover in the restoration of the site 40 years ago described the result of subsequent changes at Crofton:

> The whole (visitor) experience is lost, all that is left is a fraction of the original offering, and in the sense of national archive, a desperate desecration has been perpetrated.[5]

What are the 'drivers' of change?
1: Volunteer-dynamics

The Crofton Branch benefits from the services of a paid, part-time warden and catering manageress, but otherwise relies on volunteers. The Branch has approximately 30 active members who manage and run the entire operation, from ordering the coal to shovelling it onto the boiler, maintaining and servicing the boiler and engine, manning the café, publicity and promotion, book-keeping, liaison and representation on the Kennet & Avon Canal Trust Council, etc. – in short, the vast range of tasks essential to keep the Station pumping.

Weekend steaming rotas are organised about a year in advance. A minimum of 15 volunteers is needed on a 'steaming day', working on a rota-system driving the engines, stoking the boiler and manning the houses and grounds.

The majority of volunteers are partially or fully retired, bringing to Crofton a vast range of life-skills, most being applicable to, or adaptable to, the operation in some way.

Their commitment and generosity with their own time and money is extraordinary, and a great asset both to The Pumping Station and to the Kennet and Avon Canal Trust.

Operation and maintenance of an important historic site is a complex business, requiring wide knowledge of business, administration, human resources management, marketing, and nowadays of course, conservation. Conservation is the most esoteric of these, and therefore tends to be under-represented amongst volunteers coming from modern commercial and industrial backgrounds, or the service industries. This has sometimes manifested itself at Crofton in a lack of informed, disciplined, or coordinated action to preserve the site as a whole, or to deliver ethical conservation.

Contractors are employed where the volunteers perceive they have insufficient skills, such as for the removal and replacement of rivets in the boiler. Contractors are generally perceived to have carried out competent work, but they must be guided by the volunteers. As cash is inevitably in short supply, good conservation practice has on occasions made way for the cheapest option, and, again, a compromise in ethical standards.

Thus changes have built up little by little over the years in an *ad hoc* way, resulting in a Pumping Station now very different in appearance to that which the Trust inherited and undertook to preserve 40 years ago. Drivers of change have included worries about static visitor numbers and competition from other attractions, although mostly distant. There has been a desire to compete with other similar sites, and achieve their apparent success, whilst failing to recognise fully the historic nature and fragility of the Crofton site.

A somewhat arbitrary vision has developed of what the grounds 'should look like' (apparently influenced by the appearance of modern suburban gardens and parks) with manicured lawns suitable for marquees and awnings for summer 'village fete' events. Whilst such events are necessary for publicity and public relations, the loss of the traditional cottage gardens, the installation of modern picnic tables/benches on permanent concrete plinths, obtrusive concrete steps to the banks, and ubiquitous signs are regrettable. Most of them could have been avoided and can be removed.

There has been a well-meant but unfounded assumption that the experience of visitors will be improved by providing more steam-engines to look at. Imaginatively interpreted, the beam engines, boiler, houses and grounds would provide sufficient visitor-interest in themselves, as was demonstrated in the early days of opening.

The dynamics of decision-making in a voluntary body is sometimes by committee, on other occasions by a charismatic leader. The promotion of cohesiveness and the need to motivate appear to have led to some decisions being not in the best interests of sympathetic long-term conservation. The number of volunteers is relatively high for a site of this scale. (Kempton Park Pumping Station operates well with a similar number, about 30, but is a much larger site.) Arguably this has resulted in too many ideas and initiatives, too many 'volunteer hours' being available to implement them, and as in many volunteer organisations, there has been a lack of curatorial oversight to regulate changes. Finally there has been a perception that the threat of injury/litigation is great, a fear untempered by a knowledge of practice elsewhere. The following section looks further at this important factor.

2: 'We are afraid of being sued'
Volunteers have brought their knowledge of Health and Safety practice in modern industry into an historic building where additional dangers are evident that can best be managed by means not necessarily obvious to a newcomer to historic working machinery. As a result many volunteers have developed a fear of litigation resulting in some inappropriate intrusions on the engine, boilers, houses and grounds.

Three senior members of the Crofton Branch have insisted that 'all changes have been necessary … in compliance with the Trust's safety adviser's requirements'.

Some have forcefully expressed the view that 'more guards are needed', one quoting an occasion when he saw a father hold his child over the rocking beam and had to stop the engine as an emergency. Another had seen a mother let her toddler crawl about on window sills by an unprotected open window. These volunteers were clearly worried about litigation if an accident occurred.

Certainly, the engines and boiler at Crofton are a hazard to visitors when in operation. The law requires that risks of injury must be assessed, minimised, and the residual risks managed so as to provide as safe an environment both for visitors and attendants as is reasonably practicable. However, measures to manage the risks must take account of how the visitors are supervised, how protective measures affect their experience of the site, (for which they have paid), and the level of alteration/intrusion on the historic fabric.

Safety in perspective
1: Freedom and personal responsibility
The threat of litigation is real and feeds on an imbalance between the exercise of freedom and acting responsibility, which are inseparable in democratic society. For example, freedom to walk along a pavement implies an obligation to keep an eye open for trip hazards. In the UK it is reasonable to expect that the pavements will be fairly even, but tree-roots, vehicle-damage, and so on still cause unevenness, and one has to beware. The pavements (where they exist) in most developing countries are often far more hazardous than in the UK, and one has to exercise proportionately greater care. In neither place will warnings of the dangers be given and are not expected; I myself must exercise an appropriate level of care. Similarly, visitors to an historic site are expected to exercise an appropriate level of care.

If visitors are familiar with the type of site to be visited, eg, a ruinous castle, they may know something of the dangers to expect. If the site is unfamiliar it is reasonable to expect that visitors will be cautious, and be alert for warnings. The site-owner must take reasonable steps to make visitors safe by providing protection, information, guidance, etc. but this does not absolve a visitor from his or her duty to act sensibly.

Site-managers and visitors both expect access to be based on the reasonable assumption that visitors will act responsibly, by being fit to visit (eg, not drunk, knowingly incapable of managing steps, etc.), taking reasonable notice of information and guidance, and complying with it. They will be expected to take reasonable care of children, and not to act irrationally. If there are indications that a visitor is, or may be, incapable or may act irresponsibly, they should be supervised more closely or politely excluded from the site.

2: Total safety is an illusion
Life will never be hazard-free, nor should it be. For example, cars are amongst the greatest advances of mankind, giving the freedom of movement essential for modern life. But cars are dangerous inanimate weapons weighing around one tonne, capable of killing a human of any age when driven at only a fraction of its maximum speed.

Clearly, a strategy is needed to manage the hazards. One element of this strategy is to place the vehicles on a 'motorway' and keep pedestrians separate. But wide roads are expensive to build, use up much land, separate or disrupt communities, pollute, and are still a hazard to straying pedestrians or animals, especially when a vehicle breaks down.

Furthermore, most car journeys start and finish in areas densely populated with pedestrians, so motorways can address only part of the hazard.

Research shows that the severity of injury to a human is fundamentally dependent on the velocity of the vehicle on impact, so speed limits effectively control both the level and severity of injuries, especially in built-up areas. But at what level should the limit be set? The lower the speed, the safer for pedestrians, but the benefit of a car is that it can swiftly move people from one place to another, a facility that is rapidly lost as the speed limit is reduced. Safety on roads is undoubtedly increased by lowering speed limits, but it is generally accepted that the residual risk of injury or death in a 30 mph area is worth bearing when set against the benefit to the community as a whole of being able to complete journeys in a reasonable time. Stating such a truth so starkly is brutal, especially when faced with a grieving parent whose child has just been killed on a road by a car observing a 30 mph limit. But, as every politician knows, the costs of decisions have to be justified in relation to the benefits to individuals and to society as a whole.

As individuals and groups we are continually faced with decisions about how to minimise the risks with which we choose to live, having decided not to eliminate the dangers completely. Reducing hazards to zero is usually neither possible nor desirable in the context of the benefits to be derived by tolerating the 'residual risk' and managing it.

3: Risk aversion
The fear of taking risks is an increasing threat to our freedom. Fear is nature's way of warning us of danger and preparing us to cope with it. As such it is necessary and healthy, but the fear of injury and its consequences can unnecessarily restrict the quality of our lives. Fear can prevent children and adults from gaining experience and thus wisdom, receiving inspiration, widening horizons, increasing knowledge, learning how to recognise and manage risks, and indeed, having fun. It is these experiences that make us mature and fulfilled human beings, so our fear of risk can sadly stunt us.

The Health & Safety Executive, often accused of being 'kill-joys' encourages a positive attitude to risk:

We believe that risk management should be about practical steps to protect people from real harm and suffering - not bureaucratic back covering. If you believe some of the stories you hear, health and safety is all about stopping any activity that might possibly lead to harm.

This is not our vision of sensible health and safety – we want to save lives, not stop them. Our approach is to seek a balance between the unachievable aim of absolute safety and the kind of poor management of risk that damages lives and the economy.[6]

4: Antidotes to fear

Are there antidotes to fear, and can risk-aversion be reduced? If we do not know and understand the forces at work in our environment and what to expect we become fearful. Conversely, knowledge and understanding build confidence, so in combating fear it is important firstly to be informed.

We need to discover what causes and effects are at work, what options are open to us, what is within our control and what is beyond it. We need to look at history and learn its lessons. We need to understand what is driving the changes that are in progress, how others cope in a similar situation to ours, and where support can come from if things go wrong.

For example, borrowing money is risky, so taking out a large mortgage can be frightening. To reduce the risk, and associated fear, one needs to know the background of the lenders, the terms offered, whether there is a better deal elsewhere, what might happen to interest rates, whether one can afford the repayments, and what will happen if one becomes ill or unemployed. The risks can then be assessed more thoroughly and with the greater certainty, so reducing the fear-factor.

Secondly, it is important to know one's limitations, that is accurately to assess one's expertise and experience, to recognise the deficiencies, and plan to work within one's ability. The fear of not being able to cope, or being caught out, is then reduced.

Ultimately, when all the analysis and planning is done, one has to weigh up the risks and returns, and make a value judgement, whether borrowing money or admitting visitors to an historic site. The judgement will be an individual one, requiring not only knowledge and wisdom, but also courage. Risks cannot be eliminated entirely and it is counter-productive to attempt to do so, so we must learn to accept and manage risk.

5: Defences in law

There are two grounds on which a defence can succeed. Firstly a defendant may demonstrate that he/she was not responsible. In matters of safety it is usual for everyone to be held responsible for actions and inaction. The fact that responsibility may be held jointly or corporately, or that one did not know of the legal obligation is not a valid defence.

Additionally, or alternatively, the defendant may demonstrate that his or her actions were reasonable and not negligent. For this to succeed it is necessary to show that action was based on the best information available at the time, that advice was taken from a source believed to be credible and reliable, and that a second opinion was obtained where appropriate. It must be possible to prove that risks were assessed in relation to the reasonable responsibilities of others, and that reasonable action was taken to manage the residual risk.

Courts rule on the 'letter of the law' but its spirit was eloquently stated by Bill Callaghan, past chairman of the Health and Safety Commission in August 1966:

> My clear message is that if you are using health and safety to stop everyday activities – get a life and let others get on with theirs. But equally, if you think health and safety is a joke and that you can just ignore real risks, then try telling that to the families of the 212 workers who never went home last year. Sensible risk management is emphatically about saving lives, not stopping them.[7]

Management of visitors at Crofton Pumping Station
1: Access and Circulation

At Crofton Pumping Station over the last four decades no less than three new staircases have been installed: from ground to first floor inside the main entrance door; from top to first floor beside No 1 Engine; and from first floor to gallery level in the Boiler House.

The reasons stated are variously to improve fire safety; to provide a through-circulation route for visitors; and in the case of the ground/first floor ladder, to improve visitor access. New staircases have undoubtedly delivered these improvements, but have drastically altered the interior's ambience from that of a simple industrial building. They have crowded and intruded upon the engines, and materially altered the historic fabric.

From the author's research it is clear ladder access is almost universal in windmills, is common in watermills and is often found in other historic buildings too. Visitors accept ladders as part of the history they have paid to see. Indeed, many visitors welcome the opportunity of climbing ladders, which is outside the normal day-to-day experience of most. However, older visitors and the less mobile like to be warned in advance of the presence of such obstacles, and greatly appreciate alternative provisions to enable them to participate in the experience in some way. Rarely do they demand such provisions. There is therefore no justification on safety grounds alone for the first-floor ladder at Crofton to be 'retired' and replaced by an intrusive non-original staircase and landing.

Are the new staircases needed to provide safe circulation for visitors? Up to 10,000 visitors pay to visit Crofton Engine House annually, mostly at weekends, peaking on steaming days, when daily numbers may reach 250-300. Thus, for all but relatively few days a year the numbers of visitors touring the House at any one time are relatively low. The perceived need for a through circulation-route is therefore actually confined to relatively few days each year.

Other historic industrial sites have developed various strategies for managing widely varying numbers of visitors, including occasional overloads, whilst minimising the impact of these on the historic integrity of site itself. A common strategy is to provide alternative activities for those visitors waiting to enter a popular area, including instruction on the history or technology of the site, the provision of refreshments, etc. On very popular sites, timed entry works well from both the visitors' and the attractions' viewpoints.

2: Guarding

Access to machinery must be controlled for the protection of visitors, staff, and the machinery itself. In practical terms on historic sites this appears to have resolved into two broad approaches.

Firstly, the fortress approach. In theory, if machinery is fully guarded, as at Crofton, visitors can be left to circulate on their own in total safety. This reduces the amount of manning needed to run the site and appears to place the onus for responsible action squarely on the visitor. Unfortunately this approach suffers from several disadvantages and limitations. First, fully guarding machinery may not be practical. For example, at Crofton it is not realistic to guard the two beams effectively, leaving a significant hazard requiring other methods of management. Then a small minority of unsupervised visitors delight in challenging the fortress, usually at increased risk. Furthermore, guards may prevent access for cleaning and maintenance, often themselves forming a hazard. They are obtrusive and impoverish the experience of visitors, and of course, they themselves need maintaining. Thus, guarding on its own is rarely a completely successful option in managing visitors.

The alternative is the supervision approach. In zoos it is generally accepted that visitors' experience is impoverished by distancing them from the creatures they have come to see, and so it is with industrial sites. The risk to visitors comes not only from the machinery with which they are in close contact in a cramped industrial building but also from their own limitations. If unfamiliar with the technology or materials, they may inadvertently suffer injury, for example from touching a hot pipe or a piece of stationary machinery which moves without warning. If ill, not paying attention, distracted, or of limited perception they are liable to suffer trips, falls, or accidental contact with moving/hot machinery. If disinterested or showing off, a minority of visitors willingly put themselves in danger, and they are not exclusively young males. These risks are best managed by engaging with the visitors, firstly by attracting their attention, then by developing their interest, informing them of the hazard and maintaining their interest until they are out of the danger area.

This 'supervision approach' has several advantages. Informed, motivated visitors are less likely to suffer accidental injury through inattention or lack of knowledge. The irresponsible minority are deterred from hazardous actions, and may be excluded by a guide from the hazardous area, or site as a whole if showing signs of unreliability. Because the risk to visitors is reduced, the level of guarding can be reduced, improving the visibility of the machinery and hence visitors experience of it. Given expert and enthusiastic guiding, visitors are far more informed about the machinery they are viewing, interested in it, and satisfied by their experience. This in turn leads to recommendations to others, and improved visitor numbers. This was the experience of Crofton in its early days under Kennet and Avon Canal Trust ownership.

Conclusions

A great debt of gratitude is owed to the many volunteers who have saved a large number of industrial sites across the UK from loss in the period since the Second World War. They have brought a wealth of much-needed expertise, enthusiasm and cash to the ongoing preservation and operation of many of these sites, and achieved what governmental and national agencies would and could never have achieved on their own.

However, on historic sites with working machinery the quality of conservation and of the visitor experience is sometimes spoiled by the apparently preponderant requirements of safety legislation over those of conservation legislation, and by the fear of litigation. Clear analysis of 'historic significance', benchmarking best practice amongst sites, professional oversight, and robust conservation planning, are needed to restore a degree of balance.

At Crofton Pumping Station, the Kennet and Avon Canal Trust needs to engage in meaningful, open discussion on the future of the site, including:

Policy: definition of preservation policy and principles by the Trust Council:
- To research, record and maintain the early materials, appearance and ambience of the engines, houses, cottages and grounds
- To reinstate the late nineteenth-/early twentieth-century appearance and ambience with minimal later additions.

- To develop a long-term strategy for the site's preservation in this condition.
- To admit visitors and interpret the site, its contents and environs for their benefit.
- To assess short-, medium- and long-term risks to the engine, houses, cottages and grounds, and manage them.
- To assess risks to visitors and volunteers, reduce risks as far as is reasonable in an historic building, and manage the residual risks.

A Policy Working Group may be needed to research best practice on conservation, public access, safety, etc. on other historic industrial sites, and how to apply them at Crofton.

Statement of Significance: The cultural, technical, and aesthetic importance of every feature of the houses, machinery, and environs should be defined and written down.

Conservation Strategy: A Strategy is needed for returning the site as far as possible to the appearance it had during its pre-1968 working days. A new regime for welcoming, engaging with and supervising visitors during steaming days will be critical in facilitating a reduction in 'modern' intrusions.

Action Plan: Preparation of an action plan to implement the Strategy, including timing, funding, volunteer training, and reassessment of risks.

Curatorial oversight: Appointment of a curator with appropriate professional qualifications, e.g. AMA. Development of standards as recommended by the Museums, Libraries and Archives Council, and achievement of Accredited Museum status.

Archives: Appointment of an archivist and provision of professional training as necessary. Assembly of Crofton Branch records and those held by private individuals into one centralised archive. Interviewing of the earliest Crofton restorers and others with early memories, record of details of the appearance of the houses, grounds, boiler and engines at the time the Kennet & Avon Canal Trust took over the building in 1968 and earlier.

A new start for Crofton's third century

With the approach of the bi-centenary celebrations of the 1812 engine, the need and opportunity now exist to return the whole Crofton Pumping Station site to an appearance more faithful to its historic past which will enhance the experience of visitors at modest cost and at a manageable risk.

As conservation standards rise and visitors become more discerning, the 'suburbanisation' of this fragile site is likely to be regarded as more and more unacceptable, and could ultimately become an embarrassment to the Trust.

Alternatively, decisive leadership by the Trust Council and Crofton Branch management could reinstate a nineteenth-century ambience, and make this sensitive site an exemplar of good conservation practice of which the members will be proud.

Notes

1 *Association of Independent Museums Bulletin*, June 2007.
2 PPS5 can be downloaded free from www.communities.gov.uk.
3 www.medway.gov.uk/roch_understanding_final261009-2.pdf.
4 Neil Cossons, 'Industrial Archaeology: The Challenge of the Evidence', *The Antiquaries Journal*, 2007, vol.87.
5 Letter to the Hon. Sec., Kennet & Avon Canal Trust, October 2008, extract reproduced with permission.
6 www.hse.gov.uk.
7 *ibid*.

Index

Page numbers in **bold** refer to main entries; those in *italics* refer to illustrations.

aeroplane engine testing **89-99**
aeroplane engine types:
    Airdisco 94
    Ghost 88, 89, 91
    Gipsy 93, 94, 98
    Goblin 88
    Gyron 89, 93
    Olympus 90, 93, 96
    Pegasus 90, 93
    RB.211 96
    Spectre 92, 93
    Sprite 92, *92*
    Super Sprite 92
    T.58 93, 94
    T.64 93
    Walter 91
aeroplane types:
    Ascender 104, *105*
    Bristol.188 93
    Buccaneer 93
    Carrier 106
    Comet 88, 89, 90-1, *92*
    Concorde 90, 93, 96, 106
    DH.51 94
    DH Moth 94, 98
    Harrier 90, 93
    Kestrel 93
    Lancastrian 91
    Skeeter helicopter 93
    SR.53 92, 104, *104*
    SR.177 93
    Trident 96
    TSR.2 90, 93
    Victor 93
    Vulcan 93
    Westland Whirlwind helicopter 94
    Wright Flyer 94
    X-15 100
aerospace **88-99**, **100-107**
Air ministry 89
Airy, George 75
Albion Brewery, Bath 56

Alcaïd Omar of Alazar 36
Allanson, Gawin 53-4
Ancoats 14
Arc-et-Senans, Royal Saltworks 10
Ashby, Eric 86
Association for Industrial Archaeology 8
Astronomer Royal 73
astronomy **70-80**
Atlee, Samuel 51, 56
Attwood & White 54

Ball, Robert 77
Barbary corsairs 23
Bartley, Nehemiah 46
Beckman, Martin 26
Bedminster Distillery 46, 53, 54-5
Bélidor, Bernard Forest de 34
Biggs, James 56
Biringuccio, Vannoccio 29, *29*
Blaenavon 14, 15
Blathwayt, William 22, 30, 38
Board of Ordnance 23, 27, 30, 36, 39, 40
Board, T.H. 57
Bombay (Mumbai), India 23, 39
Bomfords, agricultural engineers 7
Booth, John 93
Boulton & Watt engines 95, 97, 98
Bourne, J.C. *19*
Bourne's Distillery, Bath 45
Bridgewater Canal 14, 71
Bridgewater Foundry 71
Bristol 59, 60, 61, 63, 65, 68
Bristol Porter Brewhouse 54
Bristol Siddeley Company 90, 93, 94, 96
British Association for the Advancement of
    Science 73, 76
Broom, Ian 110
Broseley 10
Brunel, Isambard Kingdom 8, *14*, 16-17, 18, 19, *19*, 71

Calvert, J.R. 98
Calvert, N.G. 98
Canal du Centre, Belgium 12
Canal du Midi, France 10, 11-12, *12*, *13*

Canal Lateral à la Garonne 12, *13*
Carpenter, John 77
Castel Nuovo, Naples, Italy 29
Castle, Michael 46, 55
Castle & Co 46, 57
Catherine of Braganza 22-3, 39
Cave, Thomas 54, 55, 56
Charles I 23
Charles II 23, 39
Chatham Royal Naval Dockyard 14, 16, 17
Cheese Lane Distillery, Bristol 46, 57
Chirk Aqueduct 16, *17*
Cholmley, Sir Hugh 23, 26-36, 38, 39
Coalbrookdale 10
Coker, William 61
Cole, Thomas 46, 55
Cooper, Edward 74
Copperhouse engines 97, 98
Cornish engines 95, 97-9
Cornish mining 14, 15
Cornwall 14, 15
Cossons, Neil 109, 114
Council of Engineering Institutions 83, 85
Crespi d'Alba mills, Italy 11
Cressy 7
Crofton Pumping Station 95, 97, 98, **108-123**, *115*, *120*, *121*
Cross, Allanson & Co 53-4
Cross, James 53-4, 55
Cross, Thomas 46
Cross, William 46
Croydon Airport 88
Cuba, coffee plantations in 17
Cundick, Tony 97

Daguerre, Louis 73
Dalton, John 73
Darjeeling Himalayan Railway, India 15, 18
Dartmouth, Lord 36, 37, 38
De Havilland Aircraft Company 88
De Havilland Engine Company 88, 89, *90*, 91, 93, 94
de Gomme, Bernard 28, 33
de la Rue, Warren 73, 75, 76
Derwent Valley Mills 14, 15
Distilleries, Bristol and Bath **43-57**
distilling process 43-5
Douro wine region, Portugal 17
Dyrham House 22, 38, 43

Earl of Sandwich 26
Easton, Thomas 54-5
Edwards, Samuel 54
Engelsberg Ironworks, Sweden 11
engineering education **81-7**
English Heritage 12, 13, 109-112 *passim*
European Aerospace Transporter project 104
excise duty 44, 45, 46, 55

Fairbairn, William 73
Felix Farley's Bristol Journal 46
Fielding, Henry 44
Fireside, Patricroft 71, *72*
Flamsteed, John 73
Forth rail bridge 14
Fox Talbot, William 73

Genoa 22, 27, 33-4
Ghaïlán, 'Abd Allah (Gayland) 26
Goslar, Germany 11
Grand Junction Canal 97
Great Copper Mine, Falun, Sweden 17
Great Flat Lode 15
Great Western Railway 14, 16, 18-19, *19*
Guanajuanto, Mexico 11
gunpowder engineering **22-39**

Halford Laboratory, Hatfield 89, *90*
Halley, Edmond 73-4
Hartop, Anne 71
Harvey engines 95, 98-9
Hawker Siddeley Aviation 100
Hawkins Lane Distillery, Bristol 53-4
health and safety of machinery **108-123**
Health and Safety Executive 110, 111, 118-9
Heenan & Froude 88-9, 93
Henry VIII 26
Herschel, John 73
Herschel, William 73, 74, 76
Hetling, William 56
Hodgkinson, Eaton 73
Hogarth, William 44
Hosier & Tunstall 56
Hall, William 54
Humberstone and Santa Laura mine 17

industrial preservation 108-114 *passim*
Inland Waterways Association 7
Institute of Marine Engineering 88

Institution of Civil Engineers 73
Institution of Mechanical Engineers 73, 88
Ironbridge Gorge 10, 11, *11*, 15
Ironbridge Gorge Museum Trust 11

Jacobs Well Distillery, Bristol 51-2, 56
James I 43
James II 59
James, Joshua 52-3
Jewkes, Robert 52

Kalka Shimla Railway, India 18
Karlskrona naval port and dockyard, Sweden 12
Kempton Park Pumping Station 116
Kennet & Avon Canal 95
Kennet & Avon Canal Trust 108, 111, 114-7 *passim*, 122-3
Kerr Stuart 7
Kew Bridge engines 97-9
Kinderdijk windmills, Netherlands 12

Langdon & Hetling 51
Lavaux vineyards, Switzerland 18
Lindbergh, Charles 88
Lister, R.A. & Co 7
Liverpool 14, 15
Liverpool & Manchester Railway 14, 71
Llangollen Canl 10, 12, 16, *17*
lunar observations 77, *78*, 79
Lynch, Ulysses 60-69 *passim*

Maggs, Thomas 56
Malden, Henry 86
Manchester 14, 16, 71, 72-3
Manchester Literary & Philosophical Society 72-3
Matheran Hill Railway, India 18
Matthews' (Mathews) directories 46
Maudslay, Henry 70, 74, 98
microelectronics **81-7**
Ministry of Defence 89
Ministry of Supply 10, 89
Monmouth, Duke of 59
Moore, Jonas 27, 30
Moray, Sir Robert 30
Morse, James 56
Mostar bridge, Bosnia 17
Moult, Dr 94
Mulaï, Ismâïl 37

NASA 100, 104
Nasmyth, Alexander 71
Nasmyth, James **70-80**, *71*
National Trust 8
Naylor, John 53-4
Nevis 59-69 *passim*, *60*
New Bath Directory 45-6
New Lanark Mills 14, 15
Newcomen Society 8
Niépce, Nicéphore 73
Nilgiri Rack Railway, India 18
Northgate Brewery, Bath 56

Ottoman Empire 26

Parry, Cole & Co 54-5
Parry, William 46, 54-5
Pepys, Samuel 26, 37-8
Perkins, Mr 52
Petrie, Martin 46
photography 73-4, 76-7, 79
Pickering, George *16*
Pico vineyards, Azores 17-18
Pinney, Azariah 59
Pinney, John 59
Pinney, John the Preacher 61
Pinney, John Pretor **59-69**
Pinney, Nathaniel 59
Plaster, Ron 97, 110
Pliny 27
Pontcysyllte Aqueduct 10, 12, 14, 16, *16*
Povey, Thomas 22, 30, 38
Pozzolana cement 22, 31, 33, 34
Pyne, Arthur 110

Quarrying 28-30

Rammelsberg, Germany 11
Rawlins, Stedman 65, 66
Regent's Canal Company 97
Reynolds, Nick 97, 115
Rhaetian Railway, Switzerland 17
Richards, Nicholas 65-6, 67-8
Rideau Canal, Canada 12, 17
Robertson, George 10, *11*
rocket launchers 102-4, 106
Rocket Propulsion Establishment, Westcott 91, 92
Rolls Royce Company 90, 93, 96

Rolt, L.T.C. *4*, 7-8, 9, 10-12, 13, *14*, 16, 18, *18*, 19, 43, 57, 70, 95, 97
Rolt, L.T.C., books by:
    From Sea to Sea 10
    George & Robert Stephenson 8, 95
    Inland Waterways 95
    Isambard Kingdom Brunel 8
    Landscape with Canals 8, 10
    Landscape with Figures 8, 18
    Landscape with Machines 8
    Narrow Boat 7
    The Aeronauts 95
    Thomas Telford 8
    Tools for the Job 70
    Victorian Engineering 70
Rolt Fellowship 7, 128
Roros mining settlement, Norway 10
Rosse, Lord 74
Royal Aeronautical Society 88
Royal African Company 59
Royal Astronomical Society 73, 75
Royal Institution 73, 74
Royal Observatory, Greenwich 76, 77
Rupert, Prince 30
Rutherford, Andrew (Lord Teviot) 26

St Christopher (St Kitts) 61, 63, 64, 65
Saltaire Mills 14, 15
Samuel Lynch & Co 61
San Leucio silk mill, Italy 12
Sandys, Carne and Vivian 97
Sarzanello, Liguria, Italy 29
Sayce, Samuel 56
Semmering Railway, Austria 15
Sewell copper mine, Chile 17
Sheres, Henry 32-3, *33*, 34-5, 36, 37-8, 39
Shuttle Orbiter 100-1, 106
Simmons, Roy 110
Sketchley's Directory, Bristol 46
Smeaton, John 34
Smiles, Samuel 70, 72
solar observations 75, 76, 77, 79
Space Shuttle 100-1
Spacecab 104-6, *105*
spaceplanes **100-107**
SpaceShipOne (SS1) 101
s.s. *Great Britain* 17, 19, *19*, 71
Statue of Liberty, New York 10
Stephens, James 68

Stokes Croft Distillery, Bristol 52-3
Stokes dynamometer 94, *95*
sugar estates in West Indies **59-69**, *64*

Talyllyn Railway 7-8, *18*
Tangier **22-40**, *24-5*, *32*, *37*
tarris (tarras, terres) 22, 30-4, *32*, 38, 39
telescopes 74, *75*
Telford, Thomas 8, 16
Tequila landscape, Mexico 18
Tokaj wine region, Hungary 17
Tucker, Joseph 44
Turner, J.M.W. *14*

UK *Tentative List* (of World Heritage Sites) 13, *14*, 15, 16, 19

Varberg radio station, Sweden 17
Verla board mill, Finland 12
Victoria, Queen 77
Vignoles, Charles Blacker 74-5
Vintage Sports Car Club 7
Visegrad Bridge, Bosnia 17
Vitruvius 31
Volklingen ironworks, Germany 11

Walcot Distillery, Bath 56
Wayne, Matthew 54
Weekes, Dr Thomas Pym 62, 65
Weiliczka salt mine, Poland 10
West Indies 59-69 *passim*
Whitby, near Tangier 28, 31, 36
Whitby, Yorkshire 26-7, 30, 32, 36
Wicksteed, Thomas 97
William III 44
Winwood, John 67
Worcester, Marquis of 98
Wouda pumping station, Netherlands 12
World Heritage Sites **10-21**
Wynter, Mary 22

Zollverein coalmine, Germany 17

Postscript: The Rolt Fellowship

When Tom Rolt died in 1974, the Senate of the University of Bath decided that it would be appropriate to establish a memorial to him. The previous year he had been awarded an honorary degree by the University and members of Senate strongly expressed the view that Rolt's distinctive contribution as a publicist and scholar to the engineering profession through his many books deserved to be commemorated. The form determined for this memorial was the creation of an Honorary Visiting Fellowship whereby mature engineers and other professional people with an interest in the history of technology could be encouraged to undertake projects in this field and to prepare them for publication.

Contributions towards the Fellowship Fund were gratefully received from many who had known Tom Rolt and from the institutions which he had supported, and the investment of this modest capital sum has generated sufficient interest to keep the Fund active up to the present. This has made possible the appointment of 14 Fellows, of whom six have died and the surviving eight remain actively associated with the Centre, now the History of Technology Research Unit, within the Social and Policy Sciences Department of the School of Humanities.

Those appointed held the status of Visiting Fellows at the University of Bath, and, although unsalaried, could draw up to £500 from the Fellowship Fund for the expenses of travelling, copying, printing, etc. Each Fellowship was initially intended to last for one year, but as new members were appointed the old ones retained their association with the University through the Centre for the History of Technology, and the History of Technology Seminar which the Centre has conducted very successfully over many years.

The result of this arrangement, in particular the fusion of interests between the Fellowship, the Centre and the Seminar, has been that a lively collection of scholars has been established and constantly replenished by new Rolt Fellowship appointments, with staff and research students regularly passing through it. The Fellows have made several distinguished personal contributions to scholarship, such as David Brown's sympathetic biography of William Froude, the nineteenth-century engineer and naval constructor, and they have also produced four substantial joint publications (listed below) through the Seminar.

The Rolt Fellowship scheme has thus been eminently worthwhile and has made a useful contribution to the scholarly achievements and public reputation of the University. The initial Fellowship Fund, however, has now been virtually exhausted, and it must be either replenished or terminated. Any replenishment will depend upon the future attitude of the University towards the history of technology, and that is clearly beyond the scope of the existing Fellowship. But it is hoped that the appointments of the surviving Fellows can be renewed for another three-year term, although beyond that the maintenance of the fine tradition of the excellent work of Tom Rolt must be in other hands.

Joint Publications
Engineers and Engineering, Bath: Bath University Press, 1996.
'Engineering Disasters', in *History of Technology*, vol.26, 2005.
'Case Studies in Engineering Training and Professional Education', in *Proceedings of the Institution of Civil Engineers: Engineering History and Heritage*, 162, February 2009, pp.29-37.
Landscape with Technology, Bath: Millstream Books, 2011.